Student Solutions Manual

to accompany

An Intermediate Course in Algebra
An Interactive Approach
Warr Curtis Slingerland

Bernadette Sandruck

Howard Community College

THOMSON

™

BROOKS/COLE

Australia • Canada • Mexico • Singapore • Spain • United Kingdom • United States

0-03-033331-8

For more information about our products, contact us at:
Thomson Learning Academic Resource Center
1-800-423-0563

For permission to use material from this text, contact us by:
Phone: 1-800-730-2214
Fax: 1-800-730-2215
Web: www.thomsonrights.com

Asia
Thomson Learning
5 Shenton Way #01-01
UIC Building
Singapore 068808

Australia
Nelson Thomson Learning
102 Dodds Street
South Street
South Melbourne, Victoria 3205
Australia

Canada
Nelson Thomson Learning
1120 Birchmount Road
Toronto, Ontario M1K 5G4
Canada

Europe/Middle East/South Africa
Thomson Learning
High Holborn House
50/51 Bedford Row
London WC1R 4LR
United Kingdom

Latin America
Thomson Learning
Seneca, 53
Colonia Polanco
11560 Mexico D.F.
Mexico

Spain
Paraninfo Thomson Learning
Calle/Magallanes, 25
28015 Madrid, Spain

Preface

This manual is a supplement to *An Intermediate Course in Algebra: An Interactive Approach* by Alison Warr, Catherine Curtis, and Penny Slingerland all of Mt. Hood Community College in Gresham, Oregon.

This manual is intended as a resource as you progress through the course. Please know that your active participation is the most beneficial way for you to succeed in this course. Make every effort to complete the exercises in the text on your own and do not over-utilize this manual. You will appreciate it in the long run.

Every effort has been made to prepare an error-free manual; however, mistakes may have occurred. For these I apologize. I would like to thank Paula Mikowicz and Nicole Sandruck for all of their help in preparing this document. I especially appreciate their patience with all of the last minute changes.

Bernadette Sandruck
August 2000

Table of Contents

Problem Set 1.1

1. | Independent | Dependent |
   |---|---|
   | a. A | H |
   | b. H | P |
   | c. S | T |
   | d. A | C |
   | e. M | T |

2. a. $10 + 4$; $10 + w$

   b. $4 - \dfrac{3}{20}$; $4 - \dfrac{x}{20}$

   c. $3 * 3 * 3 * 3$; 3^t

   d. $8^2 + 5 * 8$; $n^2 + 5 * n$

3. a. $5 * 6$; $5(x + 2)$

   b. $8 - 8 * 10$; $8 - 2t * 10$

   c. 3^4; $3^{\left(p/5\right)}$

   d. $0.40*8^2 + 2.5$; $0.40*(r+4)^2 + 2.5$

4. a. The second table more clearly shows the pattern.

   b. $y = 5 * 3^x$

5. a. The first table shows the pattern.

   b. $A = 200 * 1.1^t$

6. a. Answers will vary.

# of checks	Monthly Charges in $
0	$3.00
1	$3 + 0.12 (1) = $3.12
5	$3 + 0.12 (5) = $3.60
8	$3 + 0.12 (8) = $3.96
10	$3 + 0.12 (10) = $4.20
20	$3 + 0.12 (20) = $5.40

   b.
   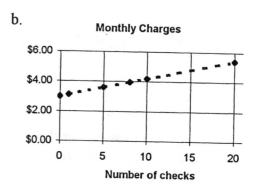
   Monthly Charges

   c. $M = 3 + 0.12n$; where n = number of checks written

   d. 4.20; 6.24; 3.00

   e. $4.92 = 3 + 0.12n$
       $1.92 = 0.12n$ → $n = 16$

       $4.90 = 3 + 0.12n$
       $1.90 = 0.12n$ → $n = 15.83$
       no; n must be a whole number

7. a. | hours rented | Cost of rental ($) |
   |---|---|
   | 4 | $10.00 |
   | 4.5 | $11.25 |
   | 5 | $12.50 |
   | 6 | $15.00 |

   b
   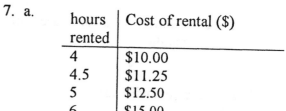
   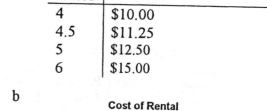
   Cost of Rental

   c. $C = \begin{cases} 10, & 0 < h \le 4 \\ 10 + 2(h-4)*1.25, & h > 4 \end{cases}$
   (fractional values of h must be rounded up to the next half hour.)

   d. $20   e. 6 hours

8. a.

# of tears	# of sheets
0	1
1	2
2	4
3	8

dependent var. = # of sheets = s
indep. var. = # of tears = t

b.

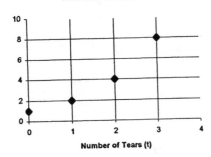

Number of sheets

c. $s = 2^t$

d – e.   Answers will vary based on height.

9. a.   Answers will vary.

Length	Width	Perimeter	Area
20	5	50 ft.	100 ft^2
15	10	50 ft.	150 ft^2
12	13	50 ft.	156 ft^2
10	15	50 ft.	150 ft^2

b.

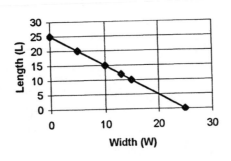

L = 25 - W

A = (25 - W) W

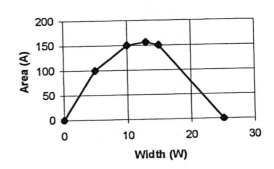

c. $2L + 2W = 50$ $L + W = 25$
     $L = 25 - W$;
     $A = L * W$
     $A = (25 - W)W$

d.   The pen should be built as a square 12.5' x 12.5'. You could use the table features to examine values between 12 ft and 13 ft. You could use the maximum feature with the area equation.

10.   31' x 62'

11. a.

n = # of years	A = amount in Account ($)
0	10,000
1	$10450
2	$10920
3	$11412

A = 10000*1.045^n

b.

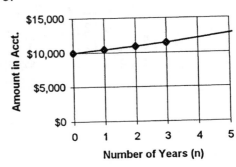

c. $A = 10,000 * 1.045^n$

d. $A = \$17721.96$

e.   Answers will vary.
     Not based on current prices of
     approximately $10,000 a year
     (unless she commutes).

12.a. $M = 500 * (0.5)^t$

  b.   15 moles are left.

13.a.

# of members	Ways to choose chairperson and secretary
2	2 * 1 = 2
3	3 * 2 = 6
4	4 * 3 = 12
n	n(n-1)

b.

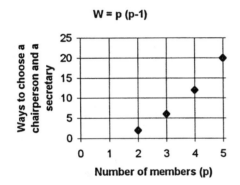

W = p (p-1)

c. $W = p(p - 1)$

d. $W = 20(19) = 380$ ways

e. $W = 49(48) = 2352$ ways

14.a.   For 10 checks choose Plan A.
        For 50 checks choose Plan B.

  b. $A = 0.25n$

  c. $B = 4 + 0.125n$

  d.   More than 32 checks.

15. $100^3 = 1,000,000 \text{ cm}^3$

16.a. $100(101) = 10100$ tiles

  b. $\dfrac{1}{2}(10100) = 5050 \text{tiles}$

c. $\dfrac{100}{2}(1+100) = 5050$

d. $\text{sum} = \dfrac{n(n+1)}{2}$

17.a.i. $a^2 + b^2 = c^2$
          $(4-1)^2 + (5-1)^2 = c^2$
          $9 + 16 = c^2$
          $c = 5$

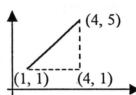

ii. $a^2 + b^2 = c^2$
      $(0--3)^2 + (7-0)^2 = c^2$
      $9 + 49 = c^2$
      $\sqrt{58} = c$

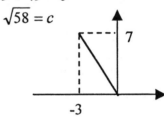

iii. $a^2 + b^2 = c^2$
       $(7-2)^2 + (4-0)^2 = c^2$
       $25 + 16 = c^2$
       $\sqrt{41} = c$

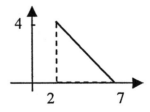

iv. $a^2 + b^2 = c^2$
      $(-3-4)^2 + (1--5)^2 = c^2$
      $49 + 36 = c^2$
      $\sqrt{85} = c$

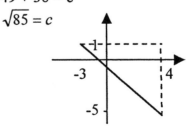

b. The distance between two points is the hypotenuse of the right triangle created by drawing a horizontal and vertical line through each endpoint. Subtract the x-values to find the length of the horizontal leg. Subtract the y-values to find the length of the vertical leg. Use the Pythagorean Theorem to solve for the hypotenuse or distance.

Distance = Square root of the sum of the square of the horizontal distance and the square of the vertical distance between the two points.

Or:
1. Square the difference between the y-values.
2. Square the difference between the x-values
3. Add the squares together
4. Take square root of the sum (from #3).

Problem Set 1.2

1. a. $y = \dfrac{25}{7}$ b. $C = 31.25$

   c. $a = 1$ d. $R = \dfrac{11}{8}$

   e. $v = \dfrac{-4}{17}$ f. $w = 2.28125$

2. a. $y = -x$; (16, -16); (9.4, -9.4); (0,0); (-10, 10)

   b. $y = 5(x + 6)$; (16,110); (9.4, 77); (0, 30); (-10, -20)

   c. $y = 5 + 6x$; (16, 101); (9.4, 61.4); (0, 5); (-10, -55)

   d. $y = 100$; (16, 100); (9.4, 100); (0, 100) (-10, 100)

3. a. $y = (3x)^2$; (4,144), (-10,900)

   b. $y = 3x^2$; (4,48), (-10, 300)

   c. $y = \dfrac{6+9+x}{3}$; (-3, 4), ((0, 5)

   d. $y = 25$; (0,25), (5,25)

   e. $y = \dfrac{2x-5}{4} + x$; (6,7.75),

       (-8, -13.25)

4. a. Pick a number. Multiply by 3. Add 6.
   b. Find the average of the input and 35, 34, and 40.
   c. Add three to the input, then square the sum
   d. Pick a number. Output is 45

5. a. No   b. Yes   c. Yes

6. a. No   b.Yes

7. Yes; for each year there is only one value for world population.

8. No, for each age there is a range of outputs for the pulse rate.

9. Yes, for each year there is one value for the output of total rainfall.

10. No, people of the same age have different running times.

11 –12. Answers will vary.

13. No, each height has more than one weight.

14. No, some number of goals scored have different shots on goal.

15.

Width	10	20	30
Length	42.5	32.5	22.5

$2W + 2L = 105$
$2L = 105 - 2W$ → $L = 52.5 - W$
Yes, for each input value for width there is a unique output for length.

16. No; a 240 lbs. person should "be careful" with 1, 2 or 3 drinks

17.a. Yes   b - c. will vary

18. Yes, each weight has a specific postal rate.

19.a. The output is the last letter of the input. E, T

   b. The letter before the first letter of the word. O, Q

   c. The output is the highest number of repeated letters. 2, 2

d. The output is the letter after the one sounded in the input. S, Z

e. It is important to know what the equation represents and adjust the window to see the full graph.

e. The output is the position in the alphabet of the first letter. 3, 2

f. The output is the number of syllables in the month. 4, 3

20. a.

x	y = 0.02x² + 1.9x - 5
-2	-8.72
0	-5
2	-1.12
4	2.92
6	7.12

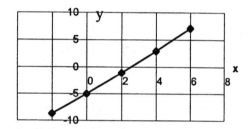

b. Yes, visual perception

c.

x	y = 0.02x² + 1.9x - 5
-125	70
-100	5
-75	-35
-50	-50
-25	-40
0	-5
25	55

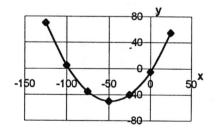

d. No, visual perception

Problem Set 2.1

1. a. $w = 4$ b. $A = 9.2$
   c. $w = -0.5$ d. $z \approx -9.714286$

2. a. $4y - 25 = 7 - x$
   $4y = 32 - x$
   $y = \dfrac{32 - x}{4}$

   b. $\dfrac{24p - 3m}{2} = 52$
   $24p - 3m = 104$
   $-3m = 104 - 24p$
   $m = -\dfrac{104}{3} + 8p \; or \; 8p - \dfrac{104}{3}$

   c. $3R(4 - K) = 2 - R$
   $12R - 3RK + R = 2$
   $13R - 3KR = 2$
   $R(13 - 3K) = 2$
   $R = \dfrac{2}{(13 - 3K)}$

   d. $\dfrac{W - T}{4T} = \dfrac{1}{8}$
   $8W - 8T = 4T$
   $8W = 12T$
   $T = \dfrac{8}{12}W = \dfrac{2}{3}W \; or \; \dfrac{2W}{3}$

3. a. $\dfrac{-4 - 0}{-2 - 2} = \dfrac{-4}{-4} = 1$

   b. $\dfrac{4 - {}^-1}{0 - 4} = \dfrac{5}{-4} = -1.25$

   c. $\dfrac{2 - {}^-6}{-10 - 5} = \dfrac{8}{-15} = \dfrac{-8}{15}$

   d. $\dfrac{25 - 125}{0 - 300} = \dfrac{-100}{-300} = \dfrac{1}{3}$

4. a. slope $= 6$ b. not linear
   c. not linear   d. $m = -5$

5. a.
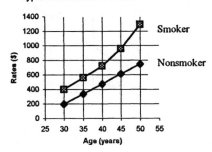
Typical Insurance Premiums

   b. Non-smokers is linear. The rate increases \$138 every 5 years; or \$27.60 for every year older.

6. a.
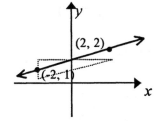

   b.
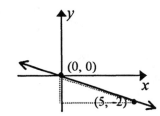

   c.
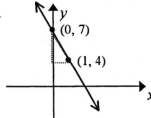

   d.

e.

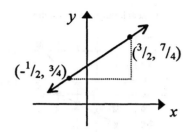

7. a.

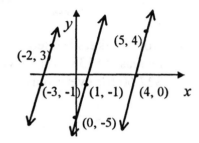

   b.   The lines are parallel.

8. a.

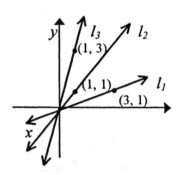

   b.   All three lines are increasing from left to right. As the slope increases the steepness of the line increases. For example, a line with m = 5 would be steeper than the graphs from part a, and a line with m = 0.5 would have less of an incline.

9.   Line A has slope = 4
        Line B has slope = 2
        Line C has slope = ¾

10. a. m = 0   b. undefined
    c. Slope of any horizontal line is 0.
    d. Slope of any vertical line is undefined.

11.   The second coordinate is (x, 0)

12. a. ≈ 325 mi.; ≈ 275 mi.
    b. -50; In 1 hr, he gets 50 mi. closer to his destination.
    c. The original distance from home is 375 miles.
    d. It takes ≈ 7.5 hrs to drive home.

13. a. 1900 calories, 3300 calories
    b. ≈ 127.27
        For each additional year between the ages of 5 and 16, a male child needs an additional 127 calories.
    c. ≈ 3700; the graph cannot be extended beyond the original data.

14.

W	$
0	420.00
1	392.50
2	365.00
3	337.50
4	310.00
5	282.50
6	255.00
7	227.50
8	200.00
9	172.50
10	145.00
11	117.50
12	90.00
13	62.50
14	35.00
15	7.50
16	-20.00

    a. Yes. m = – $27.50/ week. There will be a decrease in her account of $27.50 each week.
    b. $420 (weeks, $) She began the semester with $420.
    c. Between the 15th and 16th week, she will have withdrawn all of her money.
    d. To have enough money for the 16th week, she should withdraw $26.25 per week.

Problem Set 2.2

1. a. $x = 575.5$ b. $x = 1$

2. a. $x = \dfrac{5a+c}{b-10}$ b. $r = \dfrac{A-P}{Pt}$

   c. $F = \dfrac{PQ}{Q+P}$ d. $n = \dfrac{^-IR}{Ir-E}$

   e. $v = \pm\sqrt{\dfrac{Fr}{m}}$

3.   b & c

4. a. $y = 2x - 4$

   b. $y = -\dfrac{5}{4}x + 4$

   c. $y = -\dfrac{3}{5}x - 6$

   d. $y = \dfrac{1}{3}x + 25$

5. a. $3y = -4x + 16$

   $y = -\dfrac{4}{3}x + \dfrac{16}{3}$

   $m = -\dfrac{4}{3}; (0, \dfrac{16}{3})$

   b. $-5y = -2x + 24$

   $y = \dfrac{2x}{5} - \dfrac{24}{5}$

   $y = 0.4x - 4.8$

   $m = 0.4; (0, -4.8)$

6. a. $y = \dfrac{2}{5}x + \dfrac{12}{5}$

   $m = \dfrac{2}{5}; (0, \dfrac{12}{5})$

   b. $y = -\dfrac{9}{20}x - 6$

   $m = -\dfrac{9}{20}; (0, -6)$

7. a.i. $m = \dfrac{5 - {}^-15}{0 - {}^-30} = \dfrac{20}{30} = \dfrac{2}{3}$

   $y = \dfrac{2}{3}x + 5$

   ii. Let $y = 0 \rightarrow 0 = \dfrac{2}{3}x + 5$

   $0 = 2x + 15$

   $-15 = 2x$

   $x = -\dfrac{15}{2}; (\dfrac{-15}{2}, 0)$ or $(-7.5, 0)$

   b.i. $m = \dfrac{600 - {}^-100}{0 - 5} = \dfrac{700}{-5} = -140$

   $y = -140x + 600$

   ii. Let $y = 0 \rightarrow 0 = -140x + 600$

   $140x = 600$

   $x = \dfrac{30}{7} \rightarrow (\dfrac{30}{7}, 0)$ or $(4\,{}^2/_7, 0)$

8. a. $m = 3; (0, -2)$

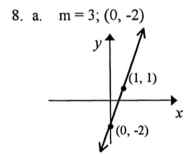

b.   m = -5; (0, 10)

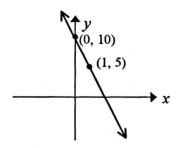

c.   m = $\frac{-3}{4}$; (0, 24)

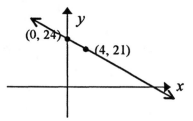

d.   m = $\frac{1}{2}$; (0, -10)

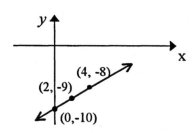

e.   m = - $\frac{4}{5}$; (0,3)

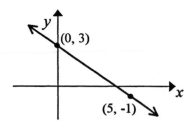

f.   m = 1; (0, 0)

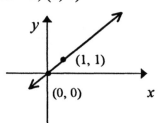

9. a.   (2.6, 0) horizontal-intercept
         (0, -13.65) vertical-intercept
   b.   (248, 0) horizontal intercept
         (0, -100) vertical intercept
   c.   (15, 0) horizontal-intercept
         (0, 25) vertical-intercept

10.a.   y = 18 – 2x

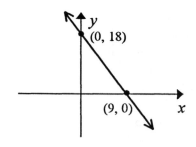

   b.   y = - 4.2x + 1.5

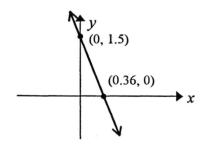

   c.   2x + 15y = 50

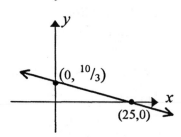

   d.   y = 10.25x – 2500

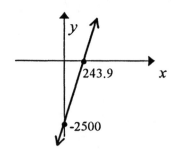

e. $2.6x + 0.25y = 24$

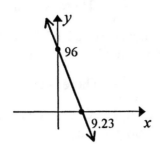

f. $\dfrac{x}{3} - y = 14$

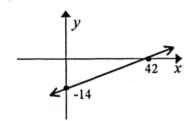

g. $3x = 7y - 8$

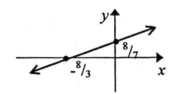

11.a. $y = 15$

b. $m = 0; y = b$

12. $x = a$; a vertical line is not a function

13.a. $m = -7$ b. $m = \sqrt{2}$

c.   not linear   d. $m = \dfrac{5}{2}$

e.   slope is undefined

f.   not linear

g. $m = \dfrac{1}{2}$ h. $m = \dfrac{5}{4}$

14.a. $m = -\dfrac{1}{2}$ b.   not linear

c. $m = 1$ d. $m = 0$
e.   not linear   f.   not linear

g. $m = \dfrac{3}{5}$ h. $m = \dfrac{3}{5}$

15.d. The slopes appear to be different because the angle of the lines changes.

e. m = 1; When you set the x and y scales to different values the slant is changed even though mathematically the slopes reduce to the same values.

16.d. The slope must be positive and the y-intercept must be positive.

17.a

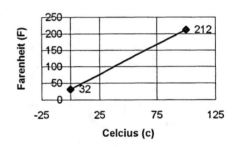

b. $F = \dfrac{9}{5}C + 32$

c. $82° \ F \approx 30° \ C$

d. $F = 60.8°$

e.   (0, 32) means that 0°C = 32°F (The freezing temperature in Celsius and Fahrenheit) (-17.8°, 0) at -17.8° C, it will be 0° F $m = \dfrac{9}{5}$; Each increase of 9°F is matched by a 5° increase in Celsius.

18.a. $F = 2C + 30$

b.   Using only freezing and melting points it is difficult to read the graph.

Comparison of Formulas

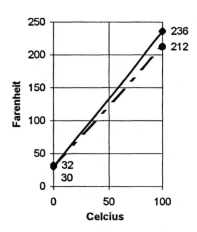

Comparison of Formulas

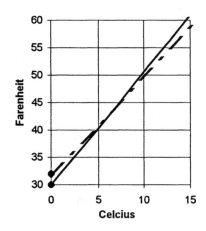

c. (10, 50)d. x > 10

19.a. B
   b. A is best up to $750 in sales.
      solution: x < $750
      B is best: $750 < x < $1125
      C is best when x > $1125
   c. Answers will vary.

20.a. window [0, 100]; [-50, 800]
      Let $Y_1 = 175 + 5.1 X$
         $Y_2 = 7.6 X$
   b. The break-even point.

c. $\dfrac{\$7.60}{1 \text{ wazoo}}$; There is an increase
   in revenue of $7.60 for each
   additional wazoo sold.
   $\dfrac{\$5.10}{1 \text{ wazoo}}$; There is an increase in
   cost of $5.10 for each additional
   wazoo produced.

d. $P = 2.5n - 175$
e. (0, -175); At 0 units produced
   (sold), you have a start up cost of
   $175.
f. break-even $\approx$ 70 wazoos (advice
   will vary)

21.a. There is a relatively consistent
      $\dfrac{\Delta y}{\Delta x}$ for the first table.

   b. (33, 2), (0, 1) → $m = \dfrac{1}{33}$

      (67, 3), (33, 2) → $m = \dfrac{1}{34}$

      (167, 6), (133, 5) → $m = \dfrac{1}{34}$

      Rounding the depth in feet to the
      nearest whole number would
      explain the slight difference in
      the slopes.

22.a. horizontal (4, 0); vertical (0, 8)
   b. Yes, for each input there is only
      one output.
   c. $D = -2t + 8$

23.a. ($\approx$ -1.75, 0); Answers may vary
      slightly
   b. (0, 1.50)   c. $m = 0.84$
      The slope will vary slightly
      depending on values chosen.
   c. $y = 0.84x + 1.50$
      (Answers will vary.)

<u>Problem Set 2.3</u>

1. a. K ≈ -0.099
   b. x ≈ 1.704
   c. m = 3
   d. x = 7

2. a. $y = \dfrac{17}{40}x + b$

   $5 = \dfrac{17}{40}(10) + b \rightarrow b = 0.75$

   y = 0.425x + 0.75

   check: (-30, -12)
   -12 = 0.425(-30) + 0.75
   -12 = -12 ✓

   b. $y = -\dfrac{500}{8}x + b$
   400 = -62.5(-2) + b → b = 275

   y = -62.5x + 275

   check: (6, -100)
   -100 = -62.5(6) + 275
   -100 = -100 ✓

3. a. y = 0.5x – 2.5
   b. y = -0.2x + 55
   c. y = -0.4x + 2
   d. y = 0.24x + 48

4. a. y = -0.5 x + 100
   b. y = 75
   c. y = x
   d. y = -x

5. a.

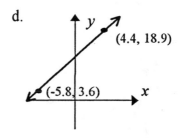

   $y = -\dfrac{3}{2}x + b$

$4 = -\dfrac{3}{2}(-1) + b \rightarrow b = 2.5$

y = -1.5x + 2.5

b.

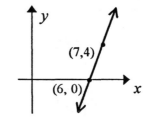

y = 4x + b
0 = 4(6) + b → b = -24
y = 4x – 24

c.

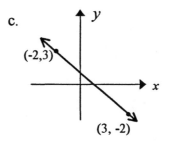

$m = \dfrac{3 - {}^-2}{{}^-2 - 3} = \dfrac{5}{-5} = -1$

y = -1x + b
3 = -1(-2) + b → b = 1
y = -x + 1

d.

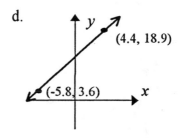

$m = \dfrac{18.9 - 3.6}{4.4 - (-5.8)} = \dfrac{15.3}{10.2} = 1.5$

3.6 = 1.5 *(-5.8) + b→ b = 12.3
y = 1.5x + 12.3

6. a.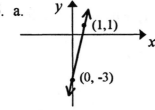

$m = 4 \rightarrow \quad y = 4x - 3$

b.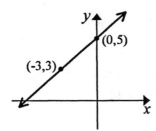

$2x - 3y = 5 \rightarrow -3y = -2x + 5$

$y = \dfrac{2}{3}x - \dfrac{5}{3}$

$m = \dfrac{2}{3} \rightarrow y = \dfrac{2}{3}x + 5$

c.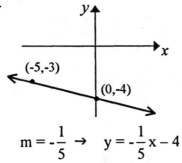

$m = -\dfrac{1}{5} \rightarrow \quad y = -\dfrac{1}{5}x - 4$

d.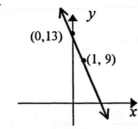

$m = -4$

$5 = -4\,(2) + b \rightarrow \quad b = 13$

$y = -4x + 13$

7. a.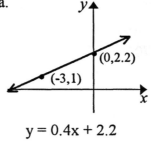

$y = 0.4x + 2.2$

b.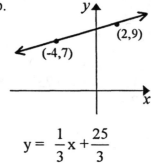

$y = \dfrac{1}{3}x + \dfrac{25}{3}$

c.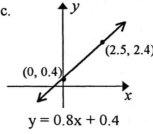

$y = 0.8x + 0.4$

d. $y = 7$

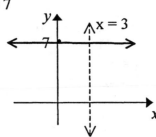

8. a. $y = 20 - 4x$

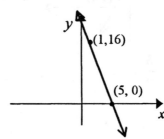

b. $y = 65x - 2500$

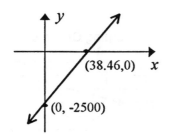

(38.46, 0) x

(0, -2500)

c. $3.2x + 0.25y = 24$

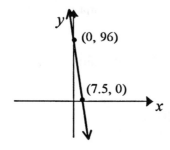

(0, 96)

(7.5, 0) x

9. a. $y = -2x$; $y = -2x + 7$;
      $y = -2x - 4$; etc.

   b. $5 = m * 4 + b$; choose values for
        m and b that make the statement
        true. $y = \frac{5}{4} x$; $y = 5$; $x = 4$

   c. $y = 5x$; $y = 5x + 2$; $y = 5x + 7$
        etc.

10. Answers will vary. These are just
    examples.
    a. $y = 3x - 4$; $y = 2x - 4$; $y = x - 4$

    b.   ex: $m = 1$; $b = -3 \rightarrow y = x - 3$
         $m = -1$; $b = 3 \rightarrow y = -x + 3$
         $m = 2$, $b = -6 \rightarrow y = 2x - 6$

    c.   Need $y = -5x + b$ where b is any
         value.

11. Answers will vary. These are just examples.

    a. $y = x$; $y = 3x$; $y = \frac{1}{2} x$

    b. $x = 2$; $x = 4$; $x = 5$

c. $y = \frac{1}{4} x$; $y = \frac{1}{4} x + 2$;

     $y = \frac{1}{4} x - 3$

12.   Many solutions. Ex: $P = 3t$

13.   To be a point you need a y value.
      $x = 5$ is the equation of a vertical
      line; therefore it is not a function.

14.a. $y = -10.5 x + 272$
    b.   272 bales
    c.   yes

15.a. $P = 12.5t + 18$

   b.

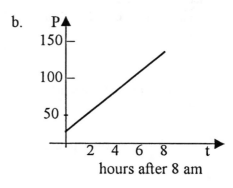

hours after 8 am

   c. $\frac{ppm}{hour}$; There is an increase in the
        pollution index of 12.5 ppm for
        each hour after 8am.
   d.   (0, 18); The pollution index is at
        18 ppm at 8am.
   e.   61.75 ppm; 211.75 ppm
        A.M. answer is reasonable (it
        falls between 8-4) but P.M. does
        not take into account the
        decrease in the pollution level
        after dark.
   f. $t \approx 7$ hrs.
        He should not be outdoors after
        3pm.

16.a. Yes; check for common slope.
m ≈ 0.151

b.
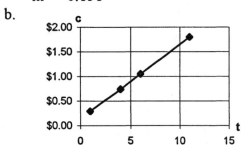

cost; cost depends on the length of call

c. y = 0.151x + 0.139 using first and last data values

d. $/min; Each additional minute costs an extra $0.15

e. The basic rate is $ 0.139 or $0.14

f. $3.16

g. 32 minutes

17. Answers will vary depending on student data.

18.a. Suburb 2 has a consistent slope of $\frac{581}{2}$.

b. $y = \frac{581}{2}x + 23112$

c. 31827 people

d. In the year 2003

19.a. 8 sq.units

b. The figure is a parallelogram because the widths are both equal and the lengths are both equal.

20.a. $M = \left(-\frac{1}{2}, \frac{9}{2}\right)$; $N = \left(5, \frac{13}{2}\right)$

b. slope of MN = $\frac{4}{11}$

slope of AB = $\frac{4}{11}$

The lines are parallel.

21. Yes; y = -2.4x - 33.8

22.a.
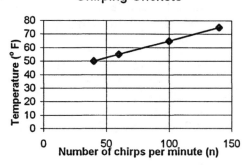

b. ≈ 60° F

c. ≈ 120° chirps/min

d. yes

e. T = 0.25n + 40

f. T = 82.5°

g. 148 chirps

h. 40° F

i. Nor reasonable, at 212° F crickets would be boiled.

23.a.& b.
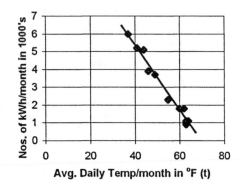

c. Answers will vary.
Using (37,6.0) and (60,1.8)
N = -0.1826t + 12.7562

d. N = $269.90

24.a. B   b. A   c. B
d. C   e. D

25. a.-b.

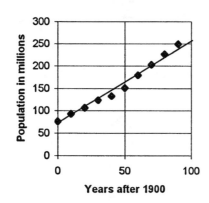

Population Growth in U.S.

c. Points used are (20, 106.0) and (80,226.5)
P = 2.0083t + 76.2
(You may get a slightly different value for "b" if you calculate using other points, rather than using the first data item.)

d. $m = \dfrac{\text{Population in millions}}{\text{Years after 1900}}$
There is an approximate increase of 2 million people in the U.S. each year since 1900.

e. P ≈ 287, 000, 000

26.b. Answers will vary depending on points chosen.
(10, 3:35.6), (30, 3:31.36)

T ≈ -0.212x + 217.72

c. $\dfrac{\text{seconds}}{\text{years after 1950}}$; Time decreases 0.212 seconds each year.

d. ≈ 3 min 26.272 sec ≈ 3:26.272

27.a.

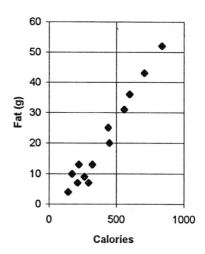

Fast Food Facts

b. Yes

c. Using (142, 4) and (840, 52)
F ≈ 0.069C − 5.765

28.

Avg. Annual % Increases from 1987 to 1994

The data is only roughly linear. A linear equation would probably not be a good predictor.

Problem Set 2.4

1. a. $2(-2) - 3(5) = -19$
         $-4 - 15 = -19$
         $-19 = -19$ ✓
         and
         $5(-2) + 2(5) = 0$
         $-10 + 10 = 0$
         $0 = 0$ ✓

    b. $5 = -3 (-2) - 1$
         $5 = 6 - 1$
         $5 = 5$ ✓
         and
         $5(-2) - 2 (5) = 11$
         $-10 - 10 = 11$
         $-20 \neq 11$ ✗

2. a.   Not a solution
    b.   Is a solution

3. a.

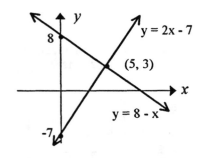

    solution: (5, 3)

    b.

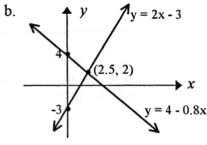

    solution: (2.5, 2)

    c.   solution: (10, -10)
    d.   solution: (6, 33)

4. There are many choices in solving these problems. Only one method is given.

    a.   Using substitution.
         $x + (2x - 5) = 7$
         $3x = 12$ → $x = 4$

         $y = 2(4) - 5$ → $y = 3$

         check: $4 + 3 = 7$
                     $7 = 7$ ✓

         solution: (4, 3)

    b.   Eliminate x.
         $4x - 5y = 9$
         $\underline{-4x - 6y = 24}$
             $-11y = 33$ → $y = -3$

         Solve for x.
         $4x - 5(-3) = 9$ → $x = -1.5$

         check: $2(-1.5) + 3(-3) = -12$
                          $-12 = -12$ ✓

         solution: (-1.5, -3)

    c.   (6, 33)   d.   (10, -10)
    e.   (5, 62.5)   f.   No solution

5.   b

6. a.   Multiply the 1st equation by 2. Begin solving by eliminating y.

    b.   Solve by substitution.

    c.   Multiply the 2nd equation by –5. Begin solving by eliminating y.

    d.   Substitute for x in the 1st equation. Be careful to change your signs.

7. a.   (4, -2)   b.   (4, 8)
    c.   (29.25, -46.5)   d.   (33.5, 10.5)
    e.   (16, 4)   f.   (-12, 0.5)
    g.   (-23.35, -15.11)
    h.   (205.71, -5.71)   i.   (2.18, -0.22)

8. a. (225, 500)   b. (-4.09, 37.73)
   d. (-128, -128)

9. a. multiply by -3
   $x + 2y - 3z = 8 \rightarrow$
   $-3x - 6y + 9z = -24$
   $\underline{3x - 4y + z = -1}$
   $-10y + 10z = -25$

   b. $x + 2y - 3z = 8$
   $\underline{-x + 10y + 4z = 9}$
   $12y + z = 17$

   c. $10y - 10z = 25$
   $\underline{120y + 10z = 170}$
   $130y \qquad = 195 \rightarrow y = 1.5$

   $-10(1.5) + 10z = -25 \rightarrow z = -1$

   d. $x + 2(1.5) - 3(-1) = 8 \rightarrow x = 2$

   e. $2 + 2(1.5) - 3(-1) = 8$
   $2 + 3 + 3 = 8$
   $8 = 8 \checkmark$

   $-2 + 10(1.5) + 4(-1) = 9$
   $-2 + 15 - 4 = 9$
   $9 = 9 \checkmark$

   $3(2) - 4(1.5) + (-1) = -1$
   $6 - 6 - 1 = -1$
   $-1 = -1 \checkmark$

   f. yes

10. a. (3, -2, 1)
    b. (0.5, -0.5, 1.5)
    c. (6, -2, 3)

11. a. $C_R = 0.75h + 10$
    $C_T = 1.22h + 2.95$

    b. (15, 21.25) For 15 hrs, both cost $21.25

    c. If you need to rent for less than 15 hrs. use TTE; for more than 15hrs. use RRE.

12. a. $x + y = 10,000$
    $0.0264x + 0.089y = 750$

    b. ($2236.42, $7763.58)

    c. No

13. $n + q = 90$
    $0.05n + 0.25q = 15.10$
    $\underline{-0.05n - 0.05q = -4.5}$
    $0.2q = 10.6 \rightarrow q = 53$

    $n = 90 - 53 = 37$

    check:
    $0.05(37) + 0.25(53) = 15.10$
    $15.10 = 15.10 \checkmark$

14. a.

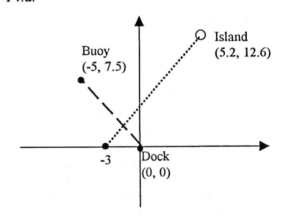

    b. $y = -1.5x; \quad y = \dfrac{63}{41}x + \dfrac{189}{41}$

    c. Pt. of collision (-1.52, 2.28)

15. a.

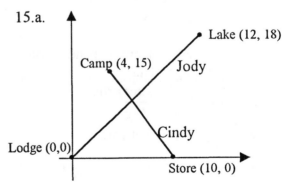

    b. Jody: $y = 1.5x$
    Cindy: $y = -2.5x + 25$

    c. (6.25, 9.375)

16.a.

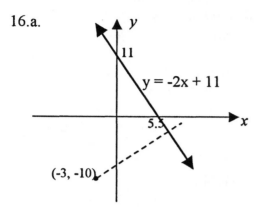

b. $\perp$ line: $y = 0.5x - 8.5$
c. $(7.8, -4.6)$
d. $d = 12.07$

17. $g = 175;\ s = 215$

18. $0.1x + 0.3y = 0.17 * 2$
can be multiplied by 100 to
eliminate the decimals
$10x + 30y = 17(2)$
    $x + y = 2 \rightarrow x = 2 - y$

Substitute the second equation
into the first equation.
$10(2 - y) + 30y = 34$
$20 - 10y + 30y = 34$
        $20y = 14 \rightarrow\ \ y = 0.7$

        $x = 2 - 0.7 = 1.3$

1.3 liters at 10%
0.7 liters at 30%

19.   Answers will vary depending on
newspaper data.
c.   slope $= \dfrac{\Delta asking\ price}{\Delta age}$
d.   original price of car.

20.a. Yes   b. No   c. $(3, 4)$

21. $(30, 18)$

Problem Set 3.1

1. a. $(-6)^5$ b. $(-4)^2$ c. $-(4)^2$

   d. $\left(\dfrac{1}{x}\right)^2$ e. $(-5)^{-2}$ f. -5^{-3}

   g. $2n^2$ h. $(2n)^{-2}$

2. $(-3)^4 = 81$; Negative 3 raised to the fourth power.
   $-3^4 = -81$; The opposite of the fourth power of three

3. a. 125   b. $\dfrac{1}{125}$ c. 1

4. a. -8 b. $\dfrac{-1}{8}$ c. -4

   d. $\dfrac{-1}{4}$ e. -1

5. a. -27 b. $\dfrac{-1}{27}$ c. 81

   d. $\dfrac{1}{81}$ e. 1

6. a. -64 b. $\dfrac{-1}{10000}$ c. -125

   d. $\dfrac{1}{36}$ e. $\dfrac{1}{49}$ f. 1

   g. $\dfrac{-1}{8}$ h. 16

7. a. $\left(\dfrac{9}{7}\right)^2$ b. $\left(\dfrac{5}{3}\right)^2$ c. $\dfrac{53}{21}$

   d. $\left(\dfrac{3}{2}\right)^3$ e. $\left(\dfrac{5}{2}\right)^4$ f. $\left(\dfrac{y}{x}\right)^4$

   g. $\left(\dfrac{b}{a}\right)^n$

8. a. $\dfrac{16}{81}$ b. $\dfrac{81}{16}$ c. 1

9. a. 36   b. $\dfrac{1}{25}$ c. $\dfrac{-27}{8}$

10. a.

x	x^2	x^3	x^4	x^5
-3	9	-27	81	-243
-2	4	-8	16	-32
-1	1	-1	1	-1
0	0	0	0	0
1	1	1	1	1
2	4	8	16	32
3	9	27	81	243

b.   **y = x**

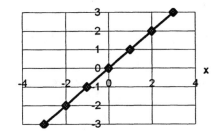

y = x²

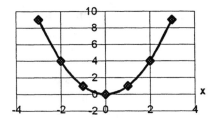

y = x³

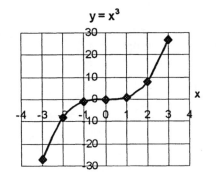

y = x⁴

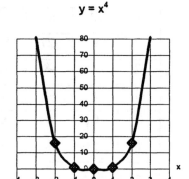

y = x⁵

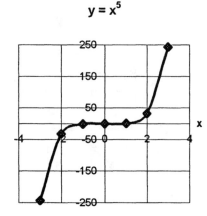

c. $y = x^6$; Similar in shape to a parabola opening upward.

$y = x^7$; looks like $y = x^3$ Function increases from left to right, passing through (0, 0).

d. $y = x^n$; when n is even is shaped like a parabola; when n is odd the graph is increasing like $y = x^3$.

11.a. 2^{-5} b. 4^{-3}

c. 5^{-3} d. 10^3

12-13. a. $9^{-2} = \dfrac{1}{81}$ b. $(-3)^{-4} = \dfrac{1}{81}$

c. $(-2)^{-3} = \dfrac{-1}{8}$

d. $(-(-10))^4 = 10,000$

e. $-4^2 = -16$ f. $-(-1)^{-2} = -1$

g. $\left(\dfrac{-1}{10}\right)^{-6} = 1,000,000$

14.a. negative; there are three negative numbers being multiplied

b. $\dfrac{-5}{64}$

15.a. negative; the x-value will be positive but y^3 will remain negative.

b. -25000

16. -109

17. $\sqrt{(-2)^2 - 4(1)(-15)}$
$= \sqrt{4 + 60} = \sqrt{64} = 8$

18.a. Positive, m^4 is positive, t^3 is negative, m is negative

b. $\dfrac{8(-3)^4(-5)^3}{16(-3)} = 1687.5$

19.a. $x^8 - 5x^8 = -4x^8$

b. $p^8 + 3\,p^8 = 4\,p^8$

c. $80\,R^3 - 60\,R^2\,T^3$

d. A^2 e. $\dfrac{1}{z^{10}}$

f. m^{14} g. $\dfrac{16}{w^8}$

20.a. $30k + 22\,k^2$ b. $\dfrac{2}{5}$

c. $x^2 + 4x + 3$ d. $3p^6$

e. $40w^2$ f. $17x^2 - \dfrac{1}{x^2}$

g. $14x^3 + \dfrac{6}{x^2}$

c. $5w^5$ d. R^3

e. x^{m+4} f. K^4m^4

g. $15w^6$ h. $112x^2k^2 - 48x^2$

21. a. $\dfrac{1}{x^5}$ b. $\dfrac{-1}{x}$

c. $4x^3$ d. $\dfrac{6}{x^5}$

e. $\dfrac{x^2}{16}$ f. $\dfrac{1}{125x^3}$

22. a. $3x^{-5}$ b. $2x^5$
    c. $4x^{-3}$ d. x^{-4}

23. a. True   b. False
    c. False   d. True
    e. False   f. False
    g. True

24. a. False   b. False
    c. False   d. False

25. a. True   b. False
    c. True   d. False

26. a. $\dfrac{1}{m^{10}}$ b. 1

c. $\dfrac{5}{x^3}$ d. $\dfrac{x^3}{5}$

e. 3   f. 1

g. $\dfrac{B^2}{A^2}$ h. $\dfrac{49Q^{-8}}{7Q^6} = \dfrac{7}{Q^{14}}$

i. $\left(\dfrac{1}{Q^{10}}\right)^2 = \dfrac{1}{Q^{20}}$

27. a. $40y^5$ b. $\dfrac{3}{64x^9}$

28. a. $4x^2 + 20xy + 25y^2$

b. $A^2 - 36B^2$

c. $y^2 - 5y + 1$

d. $2x - y$

e. $\dfrac{2H}{5} + \dfrac{16GH^5}{5}$

29. a. $-(-2)^2 * (-3)^3 = -4 * (-27) = 108$

b. $\dfrac{7(2.4^3 * -5.1)}{14(-5.1)} \approx -487.31$

c. $(-0.25)^{-2} + (-0.5)^{-3} = 8$

d. $\dfrac{1}{(-0.25)^2 + (-0.5)^3} = -16$

30. a. Answers may vary.
    b. 0.19
    c. $8.04 * 10^7$

31. a. Area $= 2(\dfrac{1}{2}bh) + lw$

$A = 2\left(\dfrac{1}{2}xh\right) + 10h$

Area $= xh + 10h$

b. Area $= \dfrac{1}{2}(A + A + B)B$

Area $= (A + 0.5B) * B$

$= AB + 0.5B^2$ or $AB + \dfrac{B^2}{2}$

Problem Set 3.2

1. a. $2^6 = 64$
   b. $3^6 = 729$
   c. $4^6 = 4096$
   d. $3^6(3^6 - 1) = 530,712$

2. 12

3. a. 10,000   b. 30,000
   c. 24,000,000

4. a. $40^3 = 64000$ b. $50^3 = 125000$

5. a.

	Brown	Blue	(mother)
Brown	BrBr	BrBl	
Blue	BlBr	BlBl	
(father)			

   b. ¾

6. a. $^1/_{16}$ b. $^1/_{16}$ c. ½

7. outcomes $= 10^4 = 10,000$
   Probability of winning $= ^1/_{10000}$

8. Assume numbers and letters can be repeated.
   a. $10*10*10 = 1000$
   b. $26^3 * 10^3 = 17,576,000$

   Proportion with TLC =
   $$\frac{1000}{17576000} = \frac{1}{17576}$$

   c. The proportion is larger because the denominator decreases in size.

9. a. 111, 112, 113, 114
      121, 122, 123, 124
      131, 132, 133, 134
      211, 212, 213, 214
      221, 222, 223, 224
      231, 232, 233, 234

   b. $\dfrac{1}{24}$ c. $\dfrac{1}{12}$ d. $\dfrac{5}{24}$

10. a.

	D	E
A	AD	AE ;
B	BD	BE
C	CD	CE

   ½

   b. AB, AC, AD, AE, BC, BD, BE, CD, CE, DE; $^2/_5$

11. a-c. done as experiment

   d. RW = 11, 12, 13, 14, 15, 16
       21, 22, 23, 24, 25, 26
       31, 32, 33, 34, 35, 36
       41, 42, 43, 44, 45, 46
       51, 52, 53, 54, 55, 56
       61, 62, 63, 64, 65, 66

   e. P(sum of 2) $= ^1/_{36}$

   P(3) $= ^2/_{36}$ P(8) $= ^5/_{36}$
   P(4) $= ^3/_{36}$ P(9) $= ^4/_{36}$
   P(5) $= ^4/_{36}$ P(10) $= ^3/_{36}$
   P(6) $= ^5/_{36}$ P(11) $= ^2/_{36}$
   P(7) $= ^6/_{36}$ P(12) $= ^1/_{36}$

12. a-c. complete experiment
    d. see sample space in 11d.

   e. P(1) $= ^1/_{36}$ P(12) $= ^4/_{36}$
   P(2) $= ^2/_{36}$ P(15) $= ^2/_{36}$
   P(3) $= ^2/_{36}$ P(16) $= ^1/_{36}$
   P(4) $= ^3/_{36}$ P(18) $= ^2/_{36}$
   P(5) $= ^2/_{36}$ P(20) $= ^2/_3$
   P(6) $= ^4/_{36}$ P(24) $= ^2/_{36}$
   P(8) $= ^2/_{36}$ P(25) $= ^1/_{36}$
   P(9) $= ^1/_{36}$ P(30) $= ^2/_{36}$
   P(10) $= ^2/_{36}$ P(36) $= ^1/_{36}$

13. a. see sample space 11d.

   b. P(7 or 11) $= ^2/_9$

   P(2, 3 or 12) $= \dfrac{1}{9}$

Chapter 1 - 3 Review

1. Answers will vary.

2. a. $w = -20$ b. $m = 10$
    c. $x = 0.5$ d. $x = 2.5$

3. Answers will vary.

4. a. $p = \dfrac{18k + 7}{5k}$

    b. $k = \dfrac{7}{5p - 18}$

    c. $y = \dfrac{x^2 - 5x}{15}$

    d. $p = \dfrac{m^2}{5m - 1}$

5. a. Yes.   b. Yes.
6. a. No   b. Yes
    c. Yes   d. No

7. a. Yes   b. No   c. Yes
8. a. Yes   b. Yes   c. No

9. a. $M = A; N = B; L = C$
    b. (0, 250) (200, 2250); $m = 10$
        Each person costs \$10 after the initial set-up fee.
    c. (0, 250); set-up fee is \$250
    d. Plan A (line M) gives the lowest cost for 75 people.

10. a, b, e, f

11. Graph A has a negative (downward) slope and a positive y-intercept.

12. a. $(x+2)(x + 3)$

    b. $6 + K^3$

    c. $c^2 + 3 * \left(\dfrac{c}{2}\right)$

    d. $6^{R\!/4}$

13. a. $-1.05°; 32.95°; 31.9°$

    b. $T = \left(\dfrac{-0.21}{100}\right)h + 34°$

    c.

    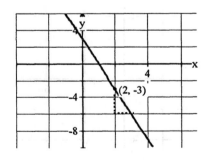

    d. $h \approx 952.4$ ft above the lodge
        6952.4 ft above sea level
    e. $\approx 23°$

14. a. $(2 + 0.5\pi)K + 2M$

    b. $K * M + \dfrac{\pi(K)^2}{4}$

15. Answers will vary.

16. a. $y = \dfrac{3}{70}x - \dfrac{37}{7}$

    b. $x = 6$

    c. $y = 12x + 30$

    d. $y = \dfrac{-10}{7}x + \dfrac{75}{7}$

17. a. $C = 12.25n + 59$
    b. 484

18. a. $y = -3x + 3$

b. $y = -x + 41$

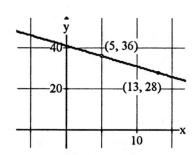

c. $y = -x - 2$

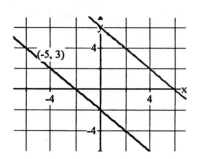

d. $y = \dfrac{1}{4}x - 3$

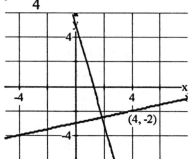

e. $x = 3$

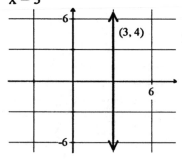

19.   Answers will vary.
      Examples:

$$y = -\frac{2}{3}x; \quad y = -\frac{2}{3}x + 2;$$

$$y = -\frac{2}{3}x - 5$$

20.   Examples: $y = -\dfrac{4}{3}x;$

$$y = -\frac{4}{3}x + 1; \quad y = -\frac{4}{3}x - 4$$

21. $m > 0 \, ; \, b < 0$

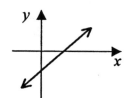

22.   ¾ minus rock $\approx 2.074 \text{ yd}^3$
      pea gravel $\approx 1.6 \text{ yd}^3$

23.a. -49 b. $-\dfrac{1}{8}$

   c. 25 d. 0.0001

   e. $\dfrac{1}{36}$ f. 8

   g. $\dfrac{9}{25}$ h. 1

24.a. $\dfrac{1}{16}$ b. $\dfrac{1}{25}$

   c. $-\dfrac{1}{16}$ d. 81

   e. -16 f. $\dfrac{1}{1000}$

   g. $-100{,}000$

25.a.　$-15x^{10}$　　b.　$40b^3 - 30b^5$

　　c.　$\dfrac{8}{k^4}$　　d.　$\dfrac{5}{x^2}$

　　e.　$\dfrac{x^2}{5}$　　f.　w^3

　　g.　$x^2 + 4x + 3$　h.　$\dfrac{1}{z^{10}}$

　　i.　$\dfrac{1}{w^8}$　　j.　$30k + 2k^2$

26.a.　$P = -0.0114a + 301.429$

　　b.

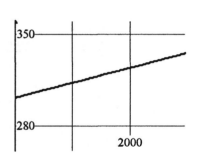

　　c.　(0, 301.429) At sea level, the horsepower $\approx$ 301.429.

　　d.　At 3000 ft above sea level, the boat has $\approx$ 267 horsepower.

　　e.　-0.0114 *or* $-\dfrac{4}{350}$ There is a decrease of 4 horsepower for each increase of 350ft. above sea level.

27.a.　(4, -2)　　b.　(2.5, 0.5)
　　c.　(15, 6)　　d.　(4, 10)
　　e.　(13, -22.5)　f.　(12,-4)

28.a.　The slopes are different. You can't tell just from a visual. Didn't adjust the viewing rectangle to see that lines are not parallel.

　　b.　(600, 211.6)

29.　Whatzit costs $3.65
　　Thingamagig costs $6.25

30.a.　(4, 1, 2)　　b.　(-2, 5 1)

31.　$w \approx 16$ weeks

32.　$s = 72$ suits; $g = 27$ gowns

33.a.　$y = \dfrac{20 - 3x}{5}$

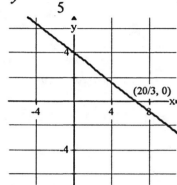

　　b.　$y = \sqrt{x + 1}$

34.　$A \approx 1011.113$ cm^2
　　$P \approx 182.8$ cm

35.a.

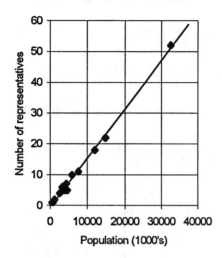

Number of Congressional Reps based on population

b. $R = (1.59*10^{-6}) P - 0.0434$

c. $1.59 \text{ E} -6 \rightarrow \dfrac{1}{627687.5}$

An increase of 627687 people gives one additional representative.

d.   8

36.a. $\dfrac{5*(-)(+)}{-2(-)} \rightarrow$ negative

b. $\dfrac{5(-5*(-3)^2)}{-2*-5} = \dfrac{5*(-5)(9)}{10} = \dfrac{-45}{2}$

37.a. ≈ 76.103
   b. ≈ -589.210
   c. ≈ -0.011

38.   Answers will vary slightly.
   90.5 min.

39. $t > 11$ years

40.a. $4p^6$ b. $40w^2$
   c. $8x - y$ d.   24

e.   12   f. $13x^2 - \dfrac{3}{x^3}$

g. $a^2 - 10ab + 24b^2$

h. $\dfrac{3}{64x^9}$ i. $22w^{12} + 2w^7$

j. $112x^2k^2 - 48x^2$

41.a. $\dfrac{1}{4}$ b.   16   c. $\dfrac{1}{16}$

   d. $\dfrac{1}{4}$ e. $\dfrac{1}{64}$

42.a. $\dfrac{1}{24}$ b. $\dfrac{1}{6}$

43.a.   162   b.   324   c.   323

<u>Problem Set 4.1</u>

1. Verify by substituting in values for the variables in the original problem and in your solution.
   a. $9A^2 - 30AB + 25B^2$
   b. $3m^3 - 2m + 1$
   c. $3x - \dfrac{12}{y}$
   d. $\dfrac{5A}{4} + \dfrac{27B^3}{4}$ or $1.25A + 6.75B^3$

2. a. b^2 b. x^{10}
   c. $x^6 + x^7$ d. $x^7 - 1$
   d. $50\pi m^2 - 3m^2$
   e. $\dfrac{5}{3}w^4$ g. $\dfrac{1}{64m^4n^6}$
   h. $\dfrac{b^4}{a^2}$ i. $\dfrac{625}{x^7y}$

3. a. $x > 3$
   b.

	x	6+5x		21	Concl.
Border Pt	3	21	=	21	✗
Solution Region	5	31	>	21	✓
Nonsolution Region	0	6	<	21	✗

4. a. $x \geq 6$
   b.

	x	-2(x-3)		3x-24	
Border Pt	6	-6	=	-6	✓
Solution Region	7	-8	<	-3	✓
Nonsol. Region	5	-4	>	-9	✗

5. a. $x \geq 3$

b. $x > \dfrac{-19}{2}$ or -9.5

c. $R \geq \dfrac{8}{5}$

d. $x > -18$

e. $w \leq \dfrac{-17}{4}$ or $-4\dfrac{1}{4}$

f. $x > \dfrac{-35}{13}$ or -2.69

g. $x > 26$

h. $C \leq 2.4$

6. Let $Y_1 = 3x + 7$; $Y_2 = 16$
   Use 2nd calc. intersect to find pt. of intersection $(3,16)$ Y_1 is greater (higher) than Y_2 when $x > 3$

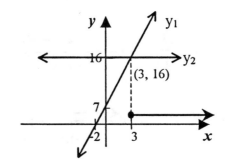

solution: $x \geq 3$

7. The line $y = -x - 9$ falls below
    $y = x + 10$ to the right of $(-9.5, .5)$
    therefore $x > -9.5$ solves the
    inequality $-x - 9 < x + 10$

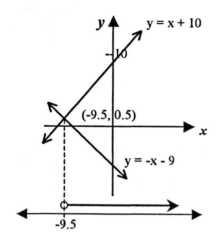

    solution: $x > -9.5$

8. a. $x < 1$

   b. $x \le 6.29$

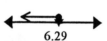

   c. $x < \dfrac{2}{3}$

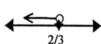

   d. $x < 1.6$

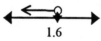

9. a. $x \ge \dfrac{11}{3}$

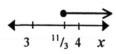

   b. $w < -1.5$

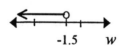

   c. $m \ge 104$

   d. $x \ge 2.75$

e. $x < 17.61$

f. $A \le 130$

g. $P < -0.677$

h. $m \ge 0.0004$

10.a. $x = 3$ b. $(3,1)$ c. $x < 3$

11.a. $x = 1$ b. $(1,-3)$ c. $x \le 1$

12.a. no solution
   b. no solution
   c. $x \in \mathfrak{R}$ (all reals)

13. $4w^2 + 62w$

14.a. Company A:
         20000 sold$\rightarrow \approx$ \$1800 in Profit
      Company B:
         20000 $\rightarrow \approx$ \$3000 in Profit; A

   b. Company A:
          45000 sold$\rightarrow \approx$ \$10000 in Profit
       Company B:
          45000$\rightarrow \approx$ \$8200 in Profit; B

   c. Less than 30000 blocks

   d. A: Using $(15, 0)$ and $(0, -50)$
       $$P = \frac{10}{3}n - 50$$
       B: Using $(5, 0)$ and $(0, -10)$
       $P = 2n - 10$

   e. $\dfrac{10}{3}n - 50 > 2n - 10$

   f. $n > 30$ thousand blocks

15.a. $45m * \dfrac{1\,min}{20m} = 2.25$ min; you

$150m * \dfrac{1\,min}{65m} = 2.31$ min; shark

Yes, you survive.

b. $\dfrac{d}{20} < 2.31$

$d < 20 * 2.31$
$d < 46.2m$; d = swimmers distance

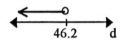

16.a.   Beth: $d = 7.5(t)$
Erik: $d = 20(t - 1.75)$

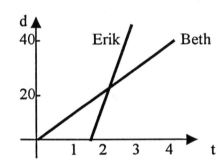

b.   7.5; The slope is the rate in miles per hour.
c.   (0,0); For Beth, at 0 time, she has traveled 0 distance.
d.   (0, -35) not appropriate.

e.   No; Yes

$t < 2$ hours 48 min.

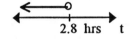

f.
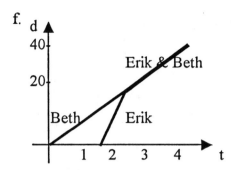

Note: you could also use t = time after they were to meet. Then
Beth: $d = 7.5(t - 0.25)$
Erik: $d = 20(t - 2)$
In part e, you have to decide if Beth was 'ahead' of Eric while she was waiting or should you adjust by 15 min.

17.a. $3.75g + 2.5s = 7000$
b. $g + s = 2500$
c.

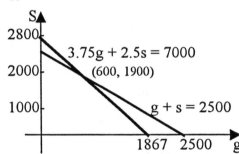

d.   yes   e.   yes

18.a. $N = \dfrac{-3}{2}p + 142.50$

  or $-1.50\,p + 142.50$

b. $p = \$95$
c. $0 \le p \le \$61.67$

19.   Output for F1 $\le$ Output for F2 when the input = 20, 25, 30, or 35. For a continuous linear function, then Input $\ge$ 20.

20.   Input $\ge$ 20

Problem Set 4.2

1. a. n^2 b. A^{10}

   c. $\dfrac{79}{r^6}$ d. $14R + 25R^2$

   e. $\dfrac{-6}{T^6}$ f. $10x^3y$

2. a.
   $\xleftarrow{\qquad}\underset{24\quad 25\quad 26\quad x}{\bullet\!\!\rightarrow}$

   b.
   $\underset{-1\quad 0\quad 1\quad k}{\circ}$

   c.
   $\underset{-5\ -4\ -3\ \ p}{\bullet\rightarrow}$

   d.
   $\underset{-11\ -10\ -9\ \ p}{\circ}$

3. a.
   $\underset{-5\quad 2\quad x}{\bullet\!-\!\bullet}$

   b.
   $\underset{2\quad 10\quad x}{\circ\quad\circ}$

   c.
   $\underset{0\quad x}{\circ\rightarrow}$

   d.
   $\underset{4\quad 10\quad x}{\circ\!-\!\circ}$

   e.
   $\underset{-2.5\quad 2.5\quad R}{\bullet\!-\!\bullet}$

4. a.
   $\underset{0\quad 8\quad x}{\circ\!-\!\circ}$

   b.
   $\underset{-4\quad -1\quad m}{\bullet\quad\bullet}$

   c.
   $\underset{k}{\longleftrightarrow}$

   d.
   $\underset{8\quad 25\quad x}{\circ\!-\!\bullet}$

   e. $x < 8$ and $x > 0$
   $\underset{0\quad 8\quad x}{\circ\!-\!\circ}$

f. $x \leq -4$ *or* $x \geq -1$
   $\underset{-4\quad -1\quad x}{\longleftarrow\!\bullet\quad\bullet\!\longrightarrow}$

5. Answers may vary.
   a. $x > -1$ b. $x \leq 3$
   c. $x \leq -5$ *or* $x \geq 10$ d. $-2 < x < 3$

6. a. $x < -20$ *or* $x > 20$
   b. $-4 \leq W \leq 5$
   c. $-10 < T < 40$
   d. $x \leq 100$ *or* $x \geq 260$

7. a. $x \leq -40$ *or* $x \geq -10$
   b. $-125 < W \leq 100$
   c. $P \leq 0$ *or* $P > 10$
   d. $-20 \leq x \leq 0$ *or* $x > 30$

8.
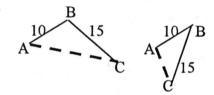

   $10 + 15 > x \ \rightarrow \ \ 25 > x$
   $15 + x > 10 \ \rightarrow \ \ x > -5$
   $10 + x > 15 \ \rightarrow \ \ x > 5$
   $5 < AC < 25$

9.   a. Make a guess.

   b. $\dfrac{86 + 82 + 73 + 78 + x}{5} \geq 80$

   c. $81 \leq x \leq 100$ $\underset{81\quad 100\quad x}{\bullet\!-\!\bullet}$

10. $x \geq -10$

11. a. $(2,0); (0,5); (-7,\sim 3)$
    b. $(-6,0); (-2,0); (2,0); (\sim 7, 0)$
    c. $x < -6$ *or* $-2 < x < 2$ *or* $x > 7$
    d. $-6 < x < -2$ *or* $2 < x < 7$

12. a. $(-2,0); (6,0)$ *or* $x = -2; 6$
    b. $x < -2$ *or* $x > 6$
    c. $-1.5 < x < 5.5$

Problem Set 4.3

1. a. $|x - 0| = 10$ *or* $|x| = 10$
   b. $|w - ^- 4| = 5$ *or* $|w + 4| = 5$
   c. $|x - 0| \geq 5$ *or* $|x| \geq 5$
   d. $|x - 7| < 4$

2. a. V   b. line
   c. line   d. line
   e. V   f. V

3. a. a V with vertex at (-9, 0)
   b. a V with vertex at (6.4, 0)
   c. a V with vertex at (0, 0)
   d. a V with vertex at (20, 0)

4. a. $x - 3 = -8$ *or* $x - 3 = 8$
      $x = -5$ *or* $x = 11$

   b. $w = -9.2$ *or* $w = -3.2$

   c. $12 - 2x = -3$ *or* $12 - 2x = 3$
      $-2x = -15$ *or* $-2x = -9$
      $x = 7.5$ *or* $x = 4.5$

   d. $q = -12$ *or* $q = 4$

   e. no solution

   f. $x \approx -9.34$
      $x \approx 37.78125$

   g. $z - 10.5 = -2.5 \rightarrow z = 8$
      or $z - 10.5 = 2.5 \rightarrow z = 13$

   h. no solution

5. a.
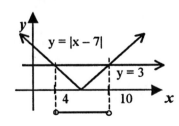

   solution: $4 < x < 10$

b.
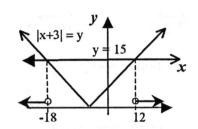

solution: $x < -18$ or $x > 12$

c. $y = |x|$
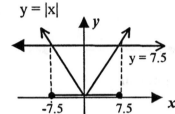

solution: $-7.5 \leq x \leq 7.5$

d.
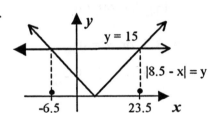

solution: $x = -6.5$ *or* $x = 23.5$

6 a.

$m \leq -18$ or $m \geq -6$

b.
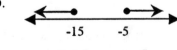

$x \leq -15$ *or* $x \geq -5$

c.

solution: $x = 9$ *or* $x = 21$

d.

solution: $q < -12$ *or* $q > 4$

11. a. $x = D$ *or* $x = F$

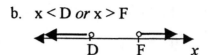

b. $x < D$ *or* $x > F$

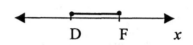

e.

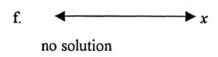

solution: all real numbers

c. $D \leq x \leq F$

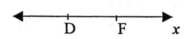

f.

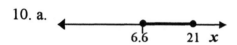

no solution

d. no solution

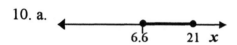

7. $15.02 \text{ oz} \leq w \leq 16.98 \text{ oz}$

8. $59.3 \text{ in.} \leq h \leq 69.7 \text{ in.}$

9. a. $x = -3.5$ *or* $x = -0.5$

    b. $x < -3.5$ *or* $x > -0.5$

10. a.

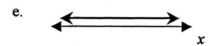

solution: $6.6 \leq x \leq 21$

b.

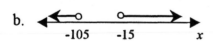

solution: $x < -105$ *or* $x > -15$

c.

no solution

Problem Set 4.4

1. a. $|x - 0| = 5 \ or \ |x| = 5$
   b. $|w - 15| = 10$
   c. $|y -^- 9| \geq 2 \ or \ |y + 9| \geq 2$
   d. $|3x -^- 7| < 4 \ or \ |3x + 7| < 4$

2. a.
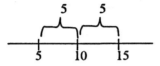
   solution: $x = 5 \ or \ x = 15$
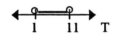

   b.
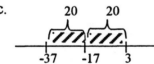
   solution: $1 < T < 11$
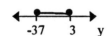

   c.
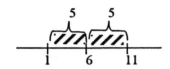
   solution: $-37 \leq y \leq 3$
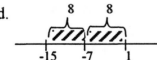

   d.
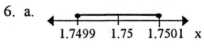 wait

3. a. solution: $w < 9.5 \ or \ w > 10.5$
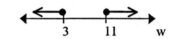

   b. solution: $x \leq -4 \ or \ x \geq -0.6$

   c. solution: $995 \leq x \leq 1015$
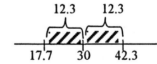

   d. All real numbers.

   e. solution: $w \leq 3 \ or \ w \geq 11$

   f. No solution.

4. 21470 hours $\leq L \leq$ 33230 hours

5.
   $17.7\% \leq P \leq 42.3\%$
   $0.177(580) = 102.66$
   $0.423(580) = 245.34$
   solution: $103 \leq$ Enroll ≤ 245

6. a.
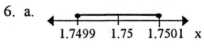
   b. $1.7499" \leq x \leq 1.7501"$
   c. $|x - 1.7500| \leq 0.0001$
   d. $|x - 1.7500| > 0.0001$

7. a.

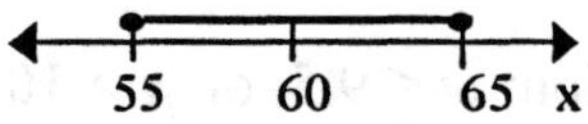

  b. $55 \text{ lbs} \le x \le 65 \text{ lbs}$

  c. $|x - 60| \le 5$

8. a. $x = -5 \ or \ x = 5 \ \rightarrow \ |x| = 5$

  b. $1 \le x \le 7 \ \rightarrow \ |x - 4| \le 3$

  c. $x \le -5 \ or \ x \ge 3 \rightarrow |x + 1| \ge 4$

  d. $P = 5 \ or \ P = 55 \rightarrow |P - 40| = 15$

  e. $-4 < t < 3 \rightarrow \left| T + \dfrac{1}{2} \right| < 3\dfrac{1}{2}$

  f. $F < 2.75 \ or \ F > 7.25$

    $|F - 5| > 2.25$

Problem Set 4.5

1.   The median score would give the
     highest grade.

2.   The mean gives the highest bonus.

3.   The mode indicates what will
     most frequently turn up.

4. a. $\bar{x} = 104.625$; $\sigma_x = 3.04$
   b.

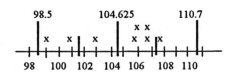

   c – d.

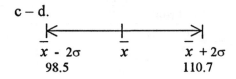

$$\bar{x} - 2\sigma \qquad \bar{x} \qquad \bar{x} + 2\sigma$$
$$98.5 \qquad\qquad\qquad 110.7$$

   e. 100%

5. a. I would expect a large standard
      deviation for both groups.
      (answers will vary)
   b. The standard deviation of a
      random group of adults should be
      greater.

6. a. $6 - 1 = 5$ games;
      $5 - 1 = 4$ games;
      Even with range, Game A has
      more variability. Not very reliable

   b. $\sigma_A \approx 1.6997$ games;
      $\sigma_B \approx 1.157$ games
      Yes; A has more variability.

7. a. $\bar{x}_{classroom} \approx 6.457$;
      $\bar{x}_{recess} \approx 5.979$. There is more
      anxiety in the classroom

   b. $\sigma_x \approx 1.955$; $\sigma_y \approx 2.209$

There is a higher variability at
recess. The data items are more
spread out from the center.

8.   75% of the scores will fall
     between 300 and 700.

9. – 10. Answers will vary

11. a.

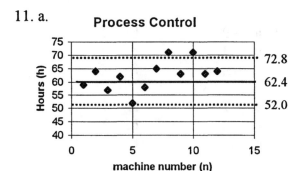

   b. $\bar{h} = 62.4$ hrs.
   c. $\sigma_h = 5.2$ hrs; $52 \leq h \leq 72.8$
   d. yes

12. a.

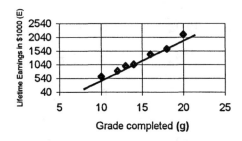

   b. Using (12, 821,000) and
      (16, 1,421,000)
      $E = 150000g - 979000$

   c.- d. Answers will vary depending
         on your equation in part b.

Problem Set 5.1

1. a. $\sqrt[5]{y}$ b. $(-27)^{\frac{1}{3}}$

   c. $5 * M^{(\frac{1}{2})}$ d. $(5m)^{\frac{1}{2}}$

   e. $-\sqrt[3]{125}$ f. $-(x)^{\frac{1}{3}}$

   g. $y^{-\frac{1}{5}}$

2. a. 9   b. -3
   c. -3 d. 2
   e. not real   f. -2

3. a. 7   b. 3
   c. 4   d. 2
   e. 25   f. $\frac{1}{5}$
   g. -6 h. $\frac{-1}{3}$

4. a. 5   b. -3
   c. not real   d. not real
   e. -1 f. not real
   g. -1 h. not real
   i. 10   j. $\frac{1}{10000}$
   k. 2

5. a. $32\sqrt[5]{x}$ b. $(M+N)^{\frac{1}{2}}$

   c. $\sqrt[5]{32x}$ d. $2\pi R^{\frac{1}{3}}$

   e. $\left(\dfrac{L}{W}\right)^{\frac{1}{2}}$ f. $-\sqrt[5]{x}$

   g. $\sqrt[5]{-x}$ h. $\dfrac{1}{\sqrt[3]{p+q}}$

   i. $\dfrac{\sqrt{w-2xy}}{5y}$

6. a. $\approx 5.2 \; or \; 5.196$
   b. $\approx 3.1 \; or \; 3.162$
   c. $\approx 25 \; or \; 27.144$

7. a. $\frac{1}{7}$ b. -1
   c. $\frac{2}{3}$ d. 3
   e. $\frac{-2}{5}$ f. not real

8. a. $9^{\frac{1}{2}}$ b. $(-27)^{\frac{1}{3}}$

   c. $(10{,}000)^{-\frac{1}{4}}$ d. $-4^{\frac{1}{2}}$

   e. $-(-1)^{\frac{1}{2}}$ f. $(-10)^{-10}$

   g. $(125)^{\frac{1}{3}}$

9. a. 3   b. -3 c. 0.1
   d. -2 e. not real

   f. $\dfrac{1}{10{,}000{,}000{,}000}$ g. 5

10. a. (Rounded to the nearest hundredth)

x	$x^{1/2}$	$x^{1/3}$	$x^{1/4}$	$x^{1/5}$
-4	not real	-1.59	not real	-1.32
-3	not real	-1.44	not real	-1.25
-2	not real	-1.26	not real	-1.15
-1	not real	-1.00	not real	-1.00
0	0.00	0.00	0.00	0.00
1	1.00	1.00	1.00	1.00
2	1.41	1.26	1.19	1.15
3	1.73	1.44	1.32	1.25
4	2.00	1.59	1.41	1.32

b.

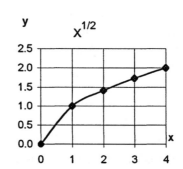

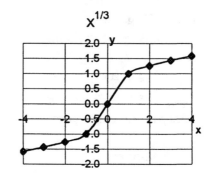

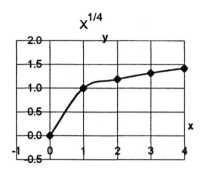

$x^{1/4}$

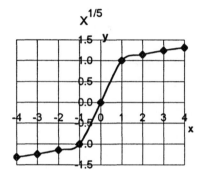

$x^{1/5}$

c. $y = x^{\frac{1}{6}}$ begins at (0, 0) increases slowly; $y = x^{\frac{1}{7}}$ looks like $y = x^{\frac{1}{3}}$ and $y = x^{\frac{1}{5}}$

d. $y = x^{\frac{1}{2}}$ when n is even; $y = x^{\frac{1}{3}}$ when n is odd

e.   when n is even, $x \geq 0$; n is odd, $x \in \Re$

11.a. $54b^{\frac{1}{4}}$ b. $H^{\frac{1}{6}}$

c. $b^0 = 1$ d. $3R^{\frac{1}{12}}$

f. $\dfrac{x}{x^{\frac{1}{2}}} + \dfrac{x^{\frac{1}{2}}}{x^{\frac{1}{2}}} = x^{\frac{1}{2}} + 1$

12.a. $\dfrac{1}{B^{\frac{1}{12}}}$ b. $-3x^3 y^{\frac{1}{6}}$

c. $1 - \dfrac{1}{y^{\frac{1}{2}}}$ d. $10m^{\frac{1}{3}}$

e. $z + 10z^{\frac{1}{2}} + 25$

13.a. $-2w^2$ b. $\dfrac{5x^4}{6}$ c.   2a

14.   B, F

15.a.   F   b.   T   c.   F   d.   T

16.a.   Neither; there is no number that can be multiplied by itself to produce a negative value.

b. $(-324)^{(-\frac{1}{2})}$ is not a real number

17.a.   Positive. The opposite of a negative is a positive number

b.   6

18.a.   neither

b. $-(-10)^{-\frac{1}{2}}$ is not a real number

19.a.   positive

b.   2.51

20.a. $G = \dfrac{6.67 * 10^{-11}(56.8)(5.98 * 10^{24})}{6.38 * 10^6}$

b. 4.2×10^9
(Answers may vary.)

c. $3.55 * 10^9$

Problem Set 5.2

1. a. $(4m)^3 = 1000$
    $4m = \sqrt[3]{1000}$
    $4m = 10$
    $m = 2.5$

    b. $5k^2 - 300 = 0$
    $5k^2 = 300$
    $k^2 = 60$
    $k \approx \pm 7.746$

    c. $10\sqrt{x-4} = 12$
    $\sqrt{x-4} = 1.2$
    $x - 4 = 1.44$
    $x = 5.44$

    d. $(120-y)^{\frac{1}{5}} = 3$
    $120 - y = 3^5$
    $120 - y = 243$
    $-123 = y$

    e. $\dfrac{\sqrt[4]{R}}{8} = 20$
    $\sqrt[4]{R} = 160$
    $R = 160^4 = 655,360,000$

2. a. $x = -1$
    b.   No solution
    c. $x = 1$ or $x = -1$
    d. $x = -1$
    e. $x = 1$
    f.   No solution

3. a. $k \geq 0$
    b. k is any real number
    c. $k \geq 0$
    d. k is any real number

4. a. $\dfrac{(2p)^5}{3} = -1$
    $(2p)^5 = -3$
    $2p = \sqrt[5]{-3}$

$p = \dfrac{\sqrt[5]{-3}}{2} \approx -0.6229$

    b. $4x^6 = 2916$
    $x^6 = 729$
    $x = \pm\sqrt[6]{729} = \pm 3$

    c. $4\sqrt{x+2} - 5 = 15$
    $4\sqrt{x+2} = 20$
    $\sqrt{x+2} = 5$
    $x + 2 = 25 \rightarrow x = 23$

    d. $12 - 6x + 12 = 3x + 1$
    $-9x = -23$
    $x = \dfrac{23}{9} \approx 2.556$

    e. $625 = (3Q-10)^4$
    $\pm 5 = 3Q - 10$
    $\dfrac{\pm 5 + 10}{3} = Q$
    $Q = 5$ *or* $Q \approx 1.667$

    f. $1000(1+x)^{12} = 2000$
    $(1+x)^{12} = 2$
    $1 + x \approx \pm 1.059463$
    $x \approx -2.059; x \approx 0.05946$

    g. $\dfrac{(x+3)}{8} = \dfrac{(5x-1)}{5}$
    $5x + 15 = 40x - 8$
    $-35x = -23$
    $x = \dfrac{23}{35} \approx 0.6571$

    h. $x^4 + 16 = 0$
    $x^4 = -16$
    No real solution

5. a. $x = -10$
    b. $x = 511$
    c. $x = \dfrac{13}{-5} = -2.6$
    d.   No solution

e. $t \approx 2.222$

f. $x \approx 4.991; x \approx 1.009$

g. $x = \pm\sqrt[2]{5} \approx \pm 2.236$

h. $x = 3$

6. a. $K = (GM)^{\frac{1}{3}}$

$(GM)^{\frac{1}{3}} = K$

$GM = K^3$

$M = \dfrac{K^3}{G}$

b. $P(1 + r)^5 = A$

$(1 + r)^5 = \dfrac{A}{P}$

$1 + r = \sqrt[5]{\dfrac{A}{P}}$

$r = -1 + \sqrt[5]{\dfrac{A}{P}}$

c. $C_p v^2 A = d$

$A = \dfrac{d}{C_p v^2}$

d. $\dfrac{4}{3}\pi r^3 = V$

$r^3 = \dfrac{3V}{4\pi}$

$r = \sqrt[3]{\dfrac{3V}{4\pi}}$

e. $2\pi\sqrt{\dfrac{L}{g}} = T$

$\sqrt{\dfrac{L}{g}} = \dfrac{T}{2\pi}$

$\dfrac{L}{g} = \dfrac{T^2}{4\pi}$

$L = \dfrac{g * T^2}{4\pi}$

7. a. $t = \pm\sqrt{\dfrac{2(x - x_0)}{a}}$

b. $M = \dfrac{(A - P_0)^3}{5.6P}$

c. $r = -k + k * \sqrt[27]{\dfrac{A}{P}}$

d. $R = \dfrac{A * i}{\left(1 - (1 + i)^{-n}\right)}$

8. a. $h \geq 7$ ft.

b. $h \approx 11.22$; Does not satisfy the minimum pitch rule unless "a" is 7 ft.

9. a. $E \approx \pm 0.05$

b. $n = \dfrac{(1.96)^2 p(1 - p)}{E^2}$

c. $n = 595$ people

d.   Assuming that it passes with greater than 50%.
$E \approx \pm 4.1$

10.   45 mph

11.a.   5%

b. ≈ 50333 people

12. $r \approx 0.56419$ ft.

13. $c = 4\sqrt{6} \approx 9.8"$

14.   diameter $\approx 12.407"$

15.   radius ≈ 2.5 ft.; diameter ≈ 5.0 ft.
$r \approx 1.898$; $d \approx 3.796$ ft.

Problem Set 5.3

1. a. $(2x)^{4/5}$ b. $\left(\sqrt{y}\right)^5$

    c. $-\sqrt[3]{x^4}$ d. $-(x)^{4/3}$

    e. $(x-3)^{3/4}$ f. $\pi\sqrt[3]{r^2}$

    g. $\pi * r^{3/2}$

2. a. $\sqrt[4]{x^5}$ b. $m^{3/2}$

    c. $25 * t^{4/3}$ d. $(25t)^{4/3}$

    e. $8\sqrt{y^5}$ f. $\left(\sqrt{8y}\right)^5$

    g. $7(cd)^{5/3}$ h. $x^{2/3}$

    i. $\sqrt[3]{\left(\dfrac{3v}{4}\right)^2}$ j. $\dfrac{-3\sqrt{x}}{y}$

3. a. $-x^{5/3}$ b. $(-y)^{5/3}$

    c. $\dfrac{n^{3/2}}{x}$ d. $\sqrt{(16x)^3}$

    e. $16\sqrt{x^3}$ f. $\left(\sqrt{-x}\right)^5$

    g. $-\sqrt{x^5}$ h. $2\pi V^{3/2}$

    i. $\left(\dfrac{c}{d}\right)^{3/2}$

4. a. $\left(5^2\right)^{3/2}=125$ b. $\left(-3^3\right)^{2/3}=9$

    c. not a real number

    d. -27 e. $10^{-2}=\dfrac{1}{100}$

    f. $\left(\dfrac{1}{2}\right)^{-2}=4$ g. $(-32)^{3/5}=-8$

5. a. $-\dfrac{1}{10000}$ b. $\dfrac{8}{27}$

c. $-\dfrac{1}{3}$ d. $9^2=81$

e. not a real number   f. 1

g. $\dfrac{1}{8}$

6. a. 1   b. -1

    c. $|x|$ d. x

7. a. $|x|$ b. $2|y|$

    c. $\dfrac{3}{B}$ d. 8x

8.   C, F

9. a. Positive. Raising any number to the 4th power is +.

    b. $(-216)^{4/3}=(-6)^4=1296$

10. a. Positive. A negative number squared will be positive.

    b. $\dfrac{1}{400}$

11. a. Neither. Square root of a negative number is not real.

    b. $y=-(-2500)^{3/2}$

12. a. Positive. The opposite of a negative is positive.

    b. ≈ 31.623

13.   b & d

14. a. $5x^{5/3}y$

    b. $\approx 3.16x^4$

    c. $45x^{7/3}y^2$

15.a   (To the nearest hundredth.)

x	$x^{1/2}$	$(x^{1/2})^3$
-4	not real	not real
-3	not real	not real
-2	not real	not real
-1	not real	not real
0	0.00	0
1	1.00	1
2	1.41	2.83
3	1.73	5.20
4	2.00	8

x	$x^{1/3}$	$(x^{1/3})^2$
-4	-1.59	2.52
-3	-1.44	2.08
-2	-1.26	1.59
-1	-1	1
0	0	0
1	1	1
2	1.26	1.59
3	1.44	2.08
4	1.59	2.52

x	$x^{1/4}$	$(x^{1/4})^5$
-4	not real	not real
-3	not real	not real
-2	not real	not real
-1	not real	not real
0	0.00	0
1	1.00	1
2	1.19	2.38
3	1.32	3.95
4	1.41	5.66

x	$x^{1/5}$	$(x^{1/5})^4$
-4	-1.32	3.03
-3	-1.25	2.41
-2	-1.15	1.74
-1	-1.00	1
0	0.00	0
1	1.00	1
2	1.15	1.74
3	1.25	2.41
4	1.32	3.03

b. $y = x^{3/2}$ and $y = x^{5/4}$ will be similar to the right side of a parabola with vertex at $(0, 0)$ opening up.

$y = x^{2/3}$ and $y = x^{5/4}$ will look like bird wings when in flight with vertex at $(0, 0)$.

c. The pairs of graphs are equivalent or look alike.

d. $y = x^{odd/even}$ will look like the positive half of a parabola opening up.

$y = x^{even/odd}$ will look like the wings of a gull or a parabola that has been split down the center and each piece flipped over the diagonal of the quadrant to face the x-axis.

e. A positive real is always appropriate input for $y = x^{n/m}$. Negative real is only appropriate if m is odd.

16.a. $x^{5/3}$ b. $\dfrac{1}{b^{7/5}}$

c. $\dfrac{5}{x^{2/3}}$ d. $6x^{5/4} - x$

e. 15.625w   f. $x^3 - 2x^{3/2}y + y^2$

g. $R^{3/2} + R$ h. $2.5t^{3/4}$

17.a. $10w^{3/2} = 80$

$w^{3/2} = 8$

$w = (8)^{2/3} = 4$

b. $\left(\sqrt{5y}\right)^3 = 4$

$\sqrt{5y} = \sqrt[3]{4}$

$5y = \left[\sqrt[3]{4}\right]^2 \approx 2.5198$

$y \approx 0.504$

c. $(B+5)^{3/4} = -2$

No real solution

d. $\sqrt{x^2 - 2.4^2} = 7.3$

$x^2 - 5.76 = 53.29$

$x^2 = 59.05$

$x \approx \pm 7.684$

18.a. $x = -3375$

   b.   No real solution

   c. $r \approx 0.175$

   d. $m \approx \pm 1.608$

   e.   No real solution

19.

0	3	6	9	24
500	$2*500$ $= 1000$	2^2*500 $= 2000$	2^3*500 $= 4000$	$2^8*500 =$ 128000

a.   **Growth of Bacteria**

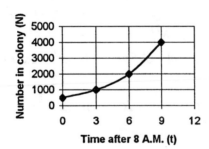

Time after 8 A.M. (t)

b. $N = 500*2^{t/3}$

c.   250 bacteria; 396 bacteria

d.   at $t \approx -27$. $N = 0.98 \to 5$ A.M. the previous day. Since this is less than one, may prefer 6 A.M.

20.

0	0.5	1	3	24
500	1000	2000	$500*2^6$ 32000	$500*2^{48}$

a.

Growth of Cholera

(graph: Number in colony (N) vs Time after 8 A.M. (t))

b. $N = 500 * 2^{2t}$

c.   250 bacteria; 125; $7.8 \to \approx 7$; round down for partial bacterium

d. $629.96 \to \approx 629$, almost 630

e.   Let $Y_2 = 1$, find intersection. $t = -4.5 \to$ back 4 hrs. 30 min. $\to$ 3:30 A.M.

21. $A = A_0 * 2^{1/k}$

22.a.   V= 161.48 hundreds of board feet

   b.

t	V
0	0
10	0.185309
20	1.048269
50	10.35911
60	16.34089
70	24.02386
100	58.6
120	92.43804
200	331.4917

c.

Old Growth

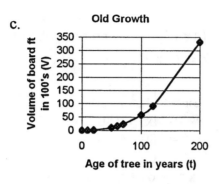

24.   ≈ 79.0 mph.

d. $5000 = 50 * 100$ board feet
Let $Y_2 = 50$ → 93.8 yrs
→ ≈ 94 yrs

23.a. $147,475.49
  b. $143,077.64
  c.   The wire can all be underwater
        meaning none is underground or
        a maximum of 3000 ft is
        underground with 500 ft
        underwater.

underground	underwater	cost
0	≈ 3,041.4	$152,069.06
500	≈ 2,549.5	$147,475.49
1000	≈ 2,061.6	$143,077.64
1500	≈ 1,581.1	$139,056.94
2000	≈ 1,118.0	$135,901.70
3000	≈ 500.0	$145,000.00

d.

Power Lines

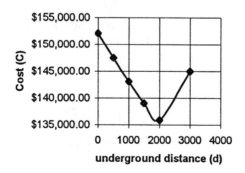

e. $C = 40x + 50\left(\sqrt{(3000-x)^2 + 500^2} \right)$

Problem Set 6.1

1. a. Linear   b. Power
    c. Exponential   d. Power
    e. Other   f. Exponential
    g. Linear   h. Other

2. a. Exponential, since ratio between each output is 6.
    b. Linear, the slope is 4.2
    c. Exponential, ratio is common for every 5 units
    d. Neither, no common difference or ratio.
    e. Exponential, common ratio of 5.
    f. Linear, slope is 1.5.

3. a. $y = 30 * 2^x$

    b. $A = 2025 * \left(\dfrac{1}{3}\right)^t$

    c. $y = 0.25 * 2^{x/10}$

    d. $C = 48.2253 * 1.2^m$

4. $B = 500 * (1.02)^{2t}$

5. a. $y = 0.5 * 6^x$

    b. $P = 10000 * 0.5^{m/15}$

    c. $C = \dfrac{42}{5}t - 146$

    d. $y = 150 * 0.6^{x/2}$

6. $B = 11 * 3^t$

7. a. $27 = 62 - x^4$
        $x^4 = 35$
        $x = \pm\sqrt[4]{35} \approx \pm 2.4323$

    b. $x = \sqrt[3]{33} \approx 3.208$

    c. $2\sqrt[5]{x-2} = 10$
        $\sqrt[5]{x-2} = 5$
        $x - 2 = 3125 \rightarrow x = 3127$

d. $x = -13$

e. $(4x+1)^{\frac{1}{2}} = -3$, no real solution

f. $r \approx 0.2011$

g. $-3x = \sqrt{4x^2 + 180}$
    $9x^2 = 4x^2 + 180$
    $5x^2 = 180$
    $x^2 = 36$
    $x = \pm 6 \rightarrow x = -6; x \neq 6$

h. $r = k\left(\sqrt[9]{\dfrac{A}{P}} - 1\right)$

8.

week	1	2	3	4	5	15
amount		$2*2$	2^3	2^4	2^5	2^{15}
(cents)	2	4	8	16	32	32768
($)	$.02	$.04	$.08	$.16	$.32	$327.68

Week 25 → ≈ $335544.32
Week 50 → ≈ $ $1.12*10^{13}$
No

9. a. 27 feet
    b.

bounces	height
0	36
1	¾ * 36 = 27
2	¾ * ¾ * 36 = 20.25
3	$(^3/_4)^3$ * 36 ≈ 15.19

c. $H = \left(\dfrac{3}{4}\right)^n *36 = 0.75^n * 36$

d. $n \approx 12.46$ when $H = 1$ →
    13 bounces for $H < 1$

10. a.

n	A
0	400
140	200
280	100
420	50
700	12.5

b. $A = 400 * \left(\dfrac{1}{2}\right)^{n/140}$

c. $n \approx 1211$ days for $A < 1$

11.a. The ratio of each successive pair of y-values is 1.03.

$y = 12690 * (1.03)^{x/2}$

b. 19771 people

c. During the year 2038 or by 2039.

12.a. Linear; $y = \dfrac{581}{2}x + 23112$

b. 31827 people

c. During the year 2003 or by 2004.

Problem Set 6.2

1. a.   Exponential   b. Power
   c.   Linear   d. Other
   e.   Exponential   f. Linear

2. a.   y-intercept (0, 1) increasing
        through (1, 2.5); horizontal
        asymptote at y = 0 as x → -∞.

   b.   y-intercept (0, 1); decreasing
        through (1, 0.15); horizontal
        asymptote at y = 0 as x → +∞

   c.   y-intercept (0, 1); decreasing
        through (-1, 3.1); horizontal
        asymptote at y = 0 as x → +∞

   d.   y-intercept (0, 1); increasing
        through (1 ,e); horizontal
        asymptote at y = 0 as x → -∞

3. a.

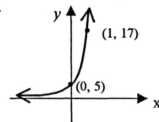

y = 5 * 3.4^x
Increasing exponential curve; y-
intercept (0, 5); horizontal
asymptote at y = 0 as x → -∞.

   b.

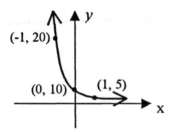

y = 10 * 0.5^x
Decreasing exponential curve
through (0, 10); horizontal
asymptote at y = 0 as x → +∞.

c.

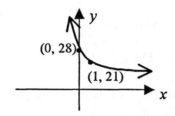

y = 28 * 0.75^x
Decreasing exponential curve
through (0, 28); horizontal
asymptote at y = 0 as x → +∞

d.

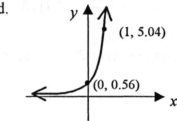

y = 0.56 * 3^{2x}
Increasing exponential curve
through (0, 0.56); horizontal
asymptote at y = 0 as x → -∞.

e.

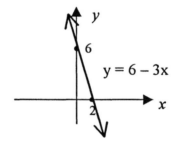

Line through (0,6) and (2, 0);
slope = -3.

f.

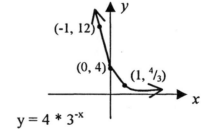

y = 4 * 3^{-x}

Decreasing exponential curve
through (0,4) horizontal
asymptote at $y = 0$ as $x \to +\infty$

4. a. Decreasing exponential curve
through (0, 12)

   b. Line with slope = -1;
decreasing through (-5, 0), (0,-5)

   c. Increasing exponential curve
through (0, 0.8)

   d. Decreasing exponential curve
through (0, 3)

   e. Increasing exponential curve
through (0, 1000)

   f. Decreasing exponential curve
through (0,0.4)

5. a. $7 * 2^{-x}$
   b. $2 * 5^{x}$
   c. $0.5 * 3^{-x}$
   d. $2 * 10^{-x}$
   e. $0.2 * e^{(-2x)}$

6. a. Decreasing through (0,7)
   b. Increasing through (0, 2)
   c. Decreasing through (0, 0.5)
   d. Decreasing through (0, 2)
   e. Decreasing through (0, 0.2)

7. a. Increasing, y-intercept (0,100)
   b. Decreasing, y-intercept (0,10)
   c. Decreasing, y-intercept (0, 50)
   d. Decreasing, y-intercept (0, 28)

8. a. $5261.90
   b. $1832.76
   c. $1826.35

9.   17yrs. 11months

10. a. Number of passenger cars

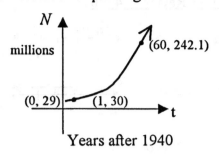

Years after 1940

    b. $\approx$ 242 million cars
    c. In 1940 $\to$ 29 cars;
58 cars $\to$ 19.5yrs or 1960
116 cars $\to$ 39.2yrs. or 1980

11. a. Predicted Population

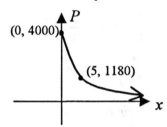

    b. 1180
    c. Yes, 13.4 years

12. a. 1800 bacteria
    b. 200 bacteria

    c. $P = 600 * (3)^{t/4}$

    d. 10.2 hours about 6:15 pm.

    e.

P
(10, 9353)
(0, 600)   (5, 2360)
t
Hours after 8 A.M.

Not all bacteria die instantly. I
allowed about 2 hrs.

13. a. 200
    b. $P = 200 * 2^{2t}$
    c. 5:30 P.M.

Chapter Review   4 – 6

1. a. $x \le -6$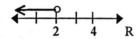
       -6   -4   x

   b. $x > -5$
       -5   -3   x

   c. $2 > R \; or \; R < 2$
       2   4   R

2. a. $x = 15$
   b. $(15, 200)$
   c. $x < 15$

3. a. $x \le -\dfrac{128}{3} \; or \; -42\dfrac{2}{3}$
       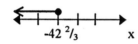
       -42 ²/₃   x

   b. $-7.5 > p \; or \; p < -7.5$
       -7.5   p

   c. $1.8 \le m \; or \; m \ge 1.8$
       -1.8   m

   d. $x \ge 8.5$
       8.5   x

4. a. $k \le 80$
   b. $x \le 6.29$
   c. $x < 3$

5. a. $y = -6x + 25$
   b. not linear
   c. $y = 25x - 144$
   d. $D = \dfrac{5}{2}t + 14$

6. a. $d = \dfrac{-65}{6}t + 480$

   b. $m = \dfrac{-65}{6}$; Every 6 hours the
       hurricane moves 65 miles closer
       to Charleston.

   c. $(0, 480)$; At 5 P.M. the hurricane
       was 480 miles away from
       Charleston.

   d. $0 = \dfrac{-65}{6}t + 480$

       Thursday, Sept. 16 at approx.
       1 P.M. → (1:18 P.M.)

   e. $180 = \dfrac{-65}{6}t + 480$

       $t = 27.7$ hrs. after 5 P.M.
       Wed., Sept.15 at 8:42 P.M.

7. a. $-2 < x < 8$
   b. $x < -2 \; or \; x > 2$
   c. $x \le -10 \; or \; x \ge 0$
   d. $20 \le x < 90$

8. $x \ge -8$

9. a. Line   b. V   c. Line
   d. V   e. V   f. Line
   g. Line

10. a. V   b. Line   c. V
    d. V   e. V   f. Line

11. a. $-6 \le W \le 10$
    b. $x \le 0 \; or \; x \ge 48$
    c. $x = -13.5 \; or \; x = -2.5$
    d. $m < -22 \; or \; m > 14$
    e. $x > 8$
    f. no solution

12. a.
        8   x

    b.
        -6   -3   R

    c.
        -4   3   k

d.

e. ![number line with open circle at 3.5 arrow right, 0 marked]

f. ![number line with open circles at -1 and 6]

13.   62 matinee tickets

14.a. $y = 4x - 7$

b. $y = \dfrac{^-7}{8}x + \dfrac{39}{8}$

15.a.   -1   b. $\dfrac{27}{8}$

c. $\dfrac{8}{27}$ d.   -8

e.   not real   f. $\dfrac{1}{16}$

g. $\dfrac{1}{16}$

16.a.   4   b.   4
c.   -4   d.   -125
e. $\dfrac{1}{25}$ f. $\dfrac{27}{8}$
g. $\dfrac{1}{81}$

17.a. $x^{7/2}$ b. $m^{5/2}$
c. $\dfrac{1}{5x}$ d. $p^{2/5}$
e. $27w^2$ f. $a^5 + 2a^{5/2}b + b^2$
g. $4x^{5/4} + x^{7/4}$ h. $2x^{7/4} + 0.5x$

18.a. $y = 150 * 2^{1/10}$
b.   (30, 1200 million barrels);
     (55, 6788.2 million barrels)
c.   66858 million barrels

19.a. $x = -7$ or $x = 2$

![number line with closed dots at -7 and 2]

b. $-7 \leq x \leq 2$

![number line segment from -7 to 2]

20.a. $4 < x < 10$
b. $x < -18$ *or* $x > 12$
c. $-7.5 \leq x \leq 7.5$
d. $x = 2$ *or* $x = 32$

21. $x < -35$

22.a. $-4.5 \leq x \leq 3$
b. $x > -15.2$
c. $x < 7.2$ *or* $x > 24$

23.a. $y = 0.25x - 9.75$
b. $y = \dfrac{^-1}{5}x + 31 = -0.2x + 31$
c. $y = -0.5x - 0.5$
d. $y = 2x + 14$
e. $y = 4$

24.a. $q \approx 173.277$ lbs. per sq. in.
b. $Q \leq 721386.87$

25.a. $m = \dfrac{4}{3p - 1}$

b. $p = \dfrac{m+4}{3m}$

c. $y = \dfrac{\sqrt{4x-3}}{5}$

d.   d. $x = \dfrac{25y^2 + 3}{4}$

e. $k = \dfrac{-3}{6a - 35}$

f.   f. $A = \pm \sqrt{B+1}$

g. $n = \dfrac{^{-}2x + 14}{x - 9}$

h. $k = \dfrac{a - 2}{5a}$

26.a.   Linear   b.   Linear
c.   Power   d.   Exponential
e.   Linear   f.   Other
g.   Power

27.a. $p = -2$ b. $m = \pm \dfrac{5}{2}$

c. $x = 7$ d. $k = 123$

e. $x = \dfrac{23}{9} \approx 2.5556$

28.a. $t \approx 0.2228$
b. $x = 7$
c. $x = \pm 6$
d.   no real solution

29.a. $A = 2000(1 + 0.08)^t$
b. $\$ 23474.16$
c. $P = \$13.24$

30.a. $x = -2373$ b. $R = -1$
c. $x = 3.5$ d. 6561
e. ± 1.7783 f. $t = 8 \; or \; t = -1$

31.a. $p = \pm \sqrt{m^2 - k^2}$

b. $r = \sqrt[5]{\dfrac{A}{2000}} - 1$

c. $B = \dfrac{1}{4}, \; A \neq 0$

d. $A = \dfrac{P}{T_0 m^3}$

e. $m = \sqrt[3]{\dfrac{P}{T_0 A}}$

f. $L = \dfrac{K^2 - B}{5}$

32.a.   Power   b.   Linear
c.   Exponential   d.   Other
e.   Other   f.   Power
g.   Linear   h.   Exponential
e.   Linear

33.a.   Increasing exponential curve through (1,8) with y-intercept at (0, 2); horizontal asymptote on the left at y = 0.
b.   Decreasing exponential curve through (1,12.5) and (0, 25) with horizontal asymptote on the right at y = 0.
c.   Decreasing exponential curve through (1, 6.75) and (0, 15) with horizontal asymptote on the right at y = 0.
d.   Increasing linear equation with slope = 3 and y-intercept at (0, –7).
e.   Increasing exponential curve through (0,1) and (1, 9) with horizontal asymptote on the left at y = 0.
f.   Decreasing exponential curve through (0, 42.5) and (1, 21.25) with horizontal asymptote on the right at y = 0.

34.a.   Let t = time after 11:30 am Friday
$d_M = 7.5(t - 4.5)$
$d_L = 6t$

b.
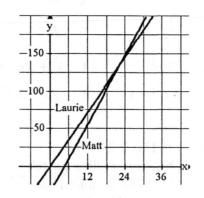

f.   12.5hrs after 11:30 am.

Matt = 60 miles
Laurie = 75 miles

g. Yes; 10 am. Sat.

h. Matt: 6pm Sat.
Laurie: 8 pm Sat.

35. a. Exponential increases by multiple of 2

b. Linear with slope = $\dfrac{12.5}{5}$

c. Common ratio = 4
Exponential

d. Neither, no common slope or common ratio.

36. a. $y = 150 * 2^x$

b. $P = 1 * 5^{m/2}$

c. $y = 4 * 3^x$

d. $D = \dfrac{5}{2}t - 22$

37. a. $C = \dfrac{-4}{3}t + 25$

b. $B = 3.2 * 2^n$

c. $P = 4096 * \left(\dfrac{5}{8}\right)^m$

d. $B = -10.4n$

38. a.

time after noon (t)	Area (A)
0	71631 cm^2
1	53723
2	40292
3	30219

b. $A = 71631 * 0.75^t$

c. $A \approx 2269$ m^2

d. 11 A.M. the next day

39. a. Decreasing exponential curve through (1, 0.42) and (0, 1) with

horizontal asymptote, y = 0, on the right side of the graph.

b. Horizontal line through y = 25.

c. Decreasing exponential curve through (0, 4) and (1, 0.8). The x-axis is the horizontal asymptote as the graph approaches + ∞.

d. Decreasing exponential curve through (0, 1) and (1, 0.8) etc.

e. Increasing exponential curve through (0, 500) and (1, 1359.1); x-axis is a horizontal asymptote as the graph approaches – ∞.

f. Decreasing through (0, 0.4) and (1, 0.04) x-axis is + ∞.

40. $\sigma_x \approx 0.533;\quad \bar{x} \approx 5.239$

41. a.

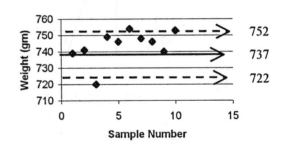

b. No (use process mean = 737)

c. $\bar{x} \approx 743.6;\ \sigma_x \approx 9.243$

<u>Problem Set 7.1</u>

1. a. Parabola, opens down; vertical
       intercept (0, 15); narrow shape;
       axis of symmetry is x = 0
   b. Parabola, opens up, vertical
       intercept (0, 20); axis of
       symmetry is x = 0
   c. Parabola, opens down; narrow
       shape; vertical intercept (0,42);
       axis symmetry is x = 0
   d. Parabola, opens up; slightly
       narrow; vertical intercept (0, -70);
       axis of symmetry is x = 0

2. a. Parabola   b. V-shape
   c. Line   d. Parabola
   e. Line   f. Line
   g. Parabola   h. Line

3. a. $y = 25 - 4x^2$

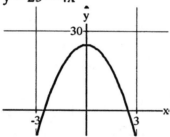

   b. $y = |2x + 15|$

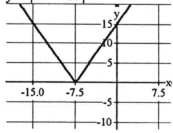

   c. $y = 5x - 2x + 12$

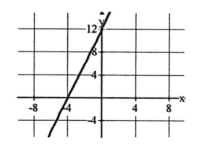

d. $y = 6x^2 + 32x$

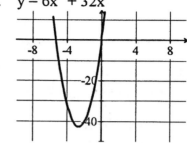

e. $x = 42$

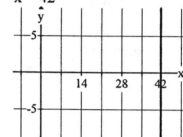

f. $y = 2x + 15$

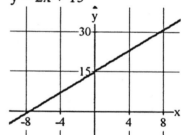

g. $y = x(5 - 0.25x)$

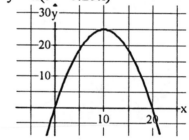

h. $y = 3(2x + 5) - x$

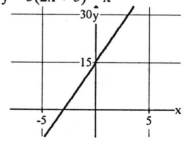

4. Many possible solutions

a.

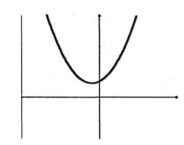

b. $y = x^2 + x + 2$

c.

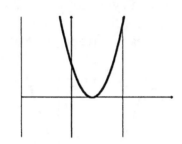

5. a. $y = -4x^2 + 484$
axis of symmetry is x = 0;
vertex = (0, 484)
= vertical intercept

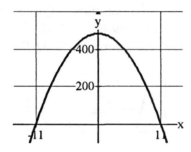

b. $y = 0.5x^2 - 8x + 17$
axis of symmetry is x = 8
vertex (8, -15)
vertical intercept (0, 17)

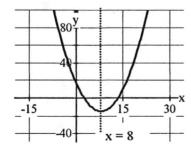

c. $y = 10x^2 + 25$
axis of symmetry is x = 0
vertex = (0, 25)

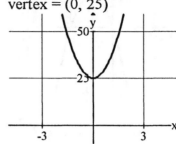

d. $y = 46x - 3x^2$
axis of symmetry is $x = 7\dfrac{2}{3}$
vertex (7.67, 176.33)
vertical intercept (0,0)

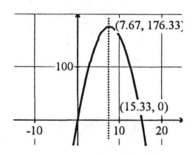

e. $y = x^2 - 15x + 60$
axis of symmetry is x = 7.5
vertex (7.5, 3.75)
vertical intercept (0, 60)

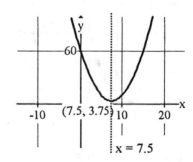

f. $y = x(10 - x)$
axis of symmetry is x = 5
vertex (5, 25)
vertical intercept (0,0)

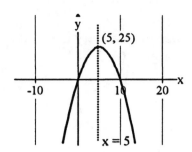

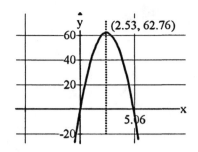

g. $y = 1280 + 100x + 5x^2$
axis of symmetry is x = -10
vertex (-10, 780)
vertical intercept (0, 1280)

j. $y = \frac{1}{2}x^2 - 20x + 180$
axis of symmetry is x = 20
vertex (20, -20)
vertical intercept (0, 180)

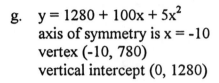

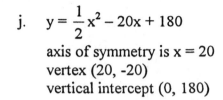

h. $y = \frac{x}{2}(18 + 2x)$
$y = 9x + x^2$
axis of symmetry is x = -4.5
vertex (-4.5, -20.25)
vertical intercept (0,0)

k. $y = 3.1x^2 + 247.2x + 5172.0$
axis of symmetry is x = -39.87
vertex (-39.87, 243.95)
vertical intercept (0, 5172.0)

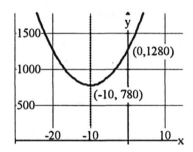

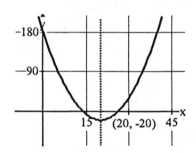

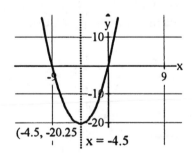

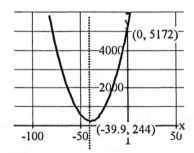

i. $y = -9.8x^2 + 49.6x$
axis of symmetry is x = 2.53
vertex (2.53, 62.76)
vertical intercept (0,0)

6.   Axis of symmetry is negative;
parabola opens up and vertical
intercept is (0, 0). ∴ F

7. a. $h = 16t^2 - 12t$

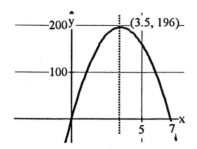

b.   maximum height 196 ft
c.   at 3.5 seconds
d.   at 7 seconds

8. a.   Area $= \left(\dfrac{50 - x}{2}\right)x$

b.

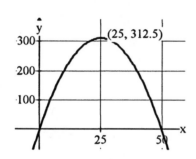

c.   vertex (25, 312.5)
dimensions 25' x 12.5'

9. a. $P = -0.51x^2 + 102x - 1500$

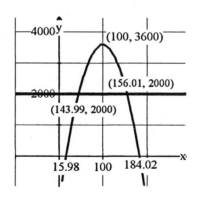

b.   -$1500, no profit
c.   100 cases
d. $3600
e. $16 < x < 184$
f. $44 < x < 156$

g.   No, maximum profit is $3600

10.a.

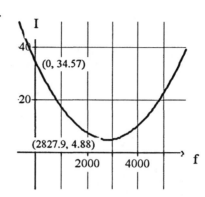

b. $x = \dfrac{-b}{2a} = \dfrac{2.1 * 10^{-2}}{2\left(3.173 * 10^{-6}\right)} \approx 2827.9$

freq. $\approx$ 2827.9 Hz

11.a.   (0, -2.5714)
$(\approx -2.8, 0)$; $(\approx 4.3, 0)$
b.   Upward; the end values are getting larger.
c.   -2.6757

12.a. $C = (0.57 + 14.50)h + 3500$
b. $I = 37h$
c. $C = \$5007$; $I = \$3700$; No
d.

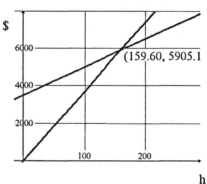

e.   Breakeven at 159.6 hrs.

13. $A = \dfrac{\pi}{2}(5 + K)^2 + 40k$

Problem Set 7.2

1. a. $-5x^2 = -80$
        $x^2 = 16$
        $x = \pm\sqrt{16} = \pm 4$

   b. $x = \dfrac{-90 \pm \sqrt{90^2 - 4(2)(1000)}}{4}$
        $x = -20; \; x = -25$

   c. $0.5x^2 - 3x = 0$
        $x(0.5x - 3) = 0$
        $x = 0; \; x = 6$

   d. $3x^2 - 2x - 8 = 0$
        $(3x + 4)(x - 2) = 0$
        $x = \dfrac{-4}{3}; \; x = 2$

   e. $4x^2 = 60$
        $x = \pm\sqrt{15} \approx \pm 3.873$

   f. $(x + 2)^2 = x^2$
        $x + 2 = \pm x$
        $x + 2 = x$ *or* $x + 2 = -x$
        $2 = 0$ $2x = -2$
        no solution; $x = -1$

2. a. $x = 15; \; x = -10$
   b. $x \approx 2.414; \; x \approx -0.414$
   c. $p = 0; \; p = 2.5$
   d. $x = {}^1/_6$
   e.   No real solution
   f. $x \approx 1041.163; \; x \approx -41.163$
   g. $x \approx 505.197; \; x \approx 2.803$
   h. $x = 3; \; x = -1$

3. a.
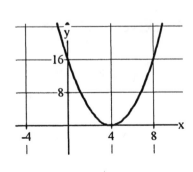

   b. $x = 4$
   c. $x = 4$; yes

4. a.   No real solution
   b.   Graph will be above the x-axis.

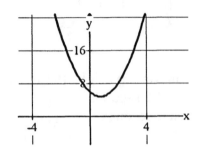

5. a. $n = 0; \; n = 3.5$
   b. $t = 0.448; \; t = 10.052$
   c. $m \approx 2.667; \; m \approx -0.667$
   d. $x = -3$
   e.   No real solution
   f. $B = \dfrac{2}{3} \approx 0.667$
   g. $w \approx 4.436; \; w \approx 0.564$
   h. $w \approx 0.785; \; w \approx 7.965$
   i. $x \approx -0.418; \; x \approx 0.718$
   j. $x \approx 197.895; \; x \approx 2.105$

6. a. $R = 23, \; R = -2$
   b. $k \approx -5.583; \; k \approx 3.583$
   c.   No real solution
   d. $x = -8.5$ or -11.5

7. a.   Other   b.   Quadratic
   c.   Linear   d.   Other
   e.   Other   f.   Quadratic
   g.   Linear   h.   Other

8. a.   Quadratic   b.   Linear
   c.   Other   d.   Quadratic
   e.   Quadratic   f.   Other
   g.   Quadratic   h.   Linear
   i.   Other   j.   Other

9. a. $\approx \pm 2.739$
   b. ≈ 4.667

c. $x = -0.5$

d. $x = 0, x = 5$

e. $x \approx -6.373, x \approx 1.373$

f. $x \approx 2.172$

g. $x = 0, x = 0.25$

h. $x = -1.6$

i. $x = 133$

j. $R \approx -7.571$

10. $x = 5, x = -6.6$

Perimeter $= 2(5+12) = 34$ inches

Area $= 5*12 = 60$ inches^2

11. $d \approx 13.05$ inches or 13 inches

12.   Sidewalk ≤ 3.8 ft.

13. $x \leq 2.36$ ft.

14. $d = 1.5m; h = 2.1m$

$$\pi r^2 h = \frac{4}{3}\pi r^3$$

$$\pi * \left(\frac{1.5}{2}\right)^2 * 2.1 = \frac{4}{3}\pi (x)^3$$

$$\left(\frac{1.5}{2}\right)^2 * 2.1 * \frac{3}{4} = x^3$$

$$0.8859375 = x^3$$

$$x \approx 0.96 \rightarrow d \approx 1.92m$$

15.   Before 7.05 seconds.

16.   approximately 7' x 36';

approximately 18' x 14'

17.a. $x \approx -12, x \approx 42$

b. $-12 < x < 42$

18.a. $x = 10$

b.   all real numbers

c. $x = 10$ (It is =, but never <.)

19. $x = -1, x = 5$

20.a.   hypotenuse $= 17 + r$

leg$_1 = 17 - r$

leg$_2 = 2\sqrt{17r}$

b. $r < 2.9$ in.

21.a.   From #16 in 4.1,

$N = -1.5p + 142.50$

$R = (-1.5p + 142.50)p$

b.   Best Price $47.50

c. ≈ 71

d. $23.22 \leq p \leq 71.78$

Problem Set 7.3

1. a. $\sqrt{m+21} = m + 1$
      $m + 21 = m^2 + 2m + 1$
      $0 = m^2 + 1m - 20$
      $0 = (m + 5)(m - 4)$
      -5 is an extraneous root
      $m = 4$

   b. $t = 1.2$
   c. $x = -2.5, x = 3$
   d. $P \approx 0.587, P \approx -0.730$

2. a. $\dfrac{3x^2}{x+5} - 1 = \dfrac{7}{x+5}$
      $3x^2 - x - 5 = 7$
      $3x^2 - x - 12 = 0$
      $x = \dfrac{1 \pm \sqrt{1 - 4(3)(-12)}}{6}$
      $x \approx -1.840, x \approx 2.174$

   b. $x \approx -1.082, x \approx 3.082$
   c. $T = \pm\dfrac{5}{7} \approx \pm 0.714$
   d. $x = -3.5$

3. a. 0   b. 1
   c. 2   d. 2

4. a. 2   b. 1
   c. 0   d. 2
   e. 0   f. 0

5.

6. a. $0 = k(x + 3)(x - 7)$
      $0 = k(x^2 - 4x - 21)$

   b. $0 = k(x - 5)(x - 20)$
      $0 = k(x^2 - 25x + 100)$

   c. $0 = k(x - \tfrac{1}{2})(x + 6)$
      $0 = k(x^2 + 5.5x - 3)$

   d. $0 = k(x - 8)^2$
      $0 = k(x^2 - 16x + 64)$

7. a. Examples: $0 = x^2 - 17x - 168$
                 $0 = 2x^2 - 34x - 336$

   b. Examples: $0 = x^2 + 77x + 1470$
                 $0 = -x^2 - 77x - 1470$

   c. Examples: $0 = x^2 + 10x + 25$
                 $0 = -3x^2 - 30x - 75$

8. a. $y = -2x^2 + 11x + 21$
   b. $y = 2.4x^2 - 16.8x - 72$

9. a. $y = \dfrac{-2}{15}x + 4$
   b. $y = -3.7x^2 + 148x - 647.5$
   c. $y = x^2 - 8x + 16$
   d. $y = 14x - 32$

10. a. $R = 93.5$ ft.   b. $P = 97.5$ psi
    c. Let $p < 20$ psi.; Answers will
       vary. 30 ft → 16.8 psi.
       25 → 13.8 psi; 20 → 10.9 psi

11. a.

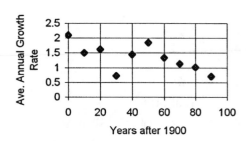

Population Growth Rate

    b. 1930, 1950
    c. Using (1900, 2.10) and
       (1940, 1.45)
       $R = -0.01625t + 2.1$
       with t = years since 1900
    d. During the year 2029 or by 2030

Problem Set 7.4

1. a. Linear   b. Quadratic
    c. Quadratic   d. Linear
    e. Quadratic   f. Linear

2. a. Quadratic   b. Quadratic
    c. Linear   d. Linear
    e. Quadratic   f. Linear
    g. Linear

3. a. $m^2 = \dfrac{35}{5P} \rightarrow m = \pm\sqrt{\dfrac{7}{P}}$

    b. $t = \dfrac{4}{p}$

    c. $12R = S^2T$

        $R = \dfrac{S^2T}{12}$

    d. $x = \dfrac{36y}{5} + 7$

    e. $m = \dfrac{3A \pm \sqrt{9A^2 - 4(AR)(4)}}{2AR}$

        $m = \dfrac{3A \pm \sqrt{9A^2 - 16AR}}{2AR}$

    f. $R = \dfrac{3AM - 4}{AM^2}$

    g. $A(RM^2 - 3M) = -4$

        $A = \dfrac{-4}{RM^2 - 3M}$

4. a. $r^2 = \dfrac{H}{8m} \rightarrow r = \pm\sqrt{\dfrac{H}{8m}}$

    b. $x = \dfrac{1 \pm \sqrt{1 - 4(R)(-Rp)}}{2R}$

        $x = \dfrac{1 \pm \sqrt{1 + 4R^2 p}}{2R}$

    c. $y = -p$ or $y = q$

    d. $2m^2 + 2Km + 5m + 5K - 9 = 0$

        $2m^2 + (2K + 5)m + (5K - 9) = 0$

        $m = \dfrac{-(2K+5) \pm \sqrt{(2K+5)^2 - 4(2)(5K-9)}}{4}$

        $m = \dfrac{-(2K+5) \pm \sqrt{4K^2 + 20K + 25 - 40K + 72}}{4}$

        $m = \dfrac{-(2K+5) \pm \sqrt{4K^2 - 20K + 97}}{4}$

5. a. $VI - I^2R = 0$

        $-I^2R = -VI$

        $R = \dfrac{VI}{I^2} = \dfrac{V}{I}$

    b. $I^2R - VI = 0$

        $I(IR - V) = 0$

        $I = 0$ or $I = \dfrac{V}{R}$

    c. $m = \dfrac{-RC \pm \sqrt{(RC)^2 - 4(LC)(1)}}{2LC}$

        $m = \dfrac{-RC \pm \sqrt{(RC)^2 - 4LC}}{2LC}$

    d. $Rt = C - S$

        $S = C - Rt$

    e. $Ma + Mb = ab$

        $Mb - ab = -Ma$

        $b(M - a) = -Ma$

        $b = \dfrac{-Ma}{M - a}$ or $\dfrac{Ma}{a - M}$

f. $Dg = g^2 + 2g + 1$

$0 = g^2 + (2 - D)g + 1$

$g = \dfrac{-(2 - D) \pm \sqrt{(2 - D)^2 - 4(1)(1)}}{2}$

$g = \dfrac{D - 2 \pm \sqrt{4 - 4D + D^2 - 4}}{2}$

$g = \dfrac{D - 2 \pm \sqrt{D^2 - 4D}}{2}$

6. $\sqrt{\dfrac{3V}{\pi h}}$

7. a. $h = \dfrac{d \pm \sqrt{d^2 - w^2}}{2}$

 b. $h \approx 0.745$, or 0.005

8. a. $x = 0$, $x = \dfrac{37}{5}$

 b. $p \approx -10.403$, $p \approx 2.4031$

 c. $k = 12.6$

 d. $y \approx 0.52519$ or $y \approx 3.8081$

 e. $r \approx 1.3561$, $r \approx -1.1061$

9. a. $x \approx 1.2749$, $x \approx -6.2749$

 b. $Q \approx 16.483$; $Q \approx 1.5167$

 c. $m = 7$

 d. $B = 2$

 e. $x = \pm 6$

10. $r = 60$ mph

11. $L \approx 26$ ft.

 dimensions: 16 ft. x 26 ft.

12.a. The graph is a circle with a radius 5 and center at (0,0).

 b. Circle with radius 7 and centered at (0,0).

 c. Circle with radius $\sqrt{7}$ and centered at (0,0)

13. Circle with radius 5, centered at (3,0)

<u>Problem Set 7.5</u>

1. a. (2.193, 5.193); (-3.193, -0.193)
   b. (-6.538, -15.538); (9.788, -7.106)
   c. No real solution
   d. (10, 2)

2. For each inequality, let Y_1 = left side and Y_2 = right side. Find the pts. of intersection.

   a.

   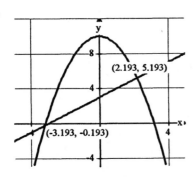

   solution: $-3.193 < x < 2.193$

   b.

   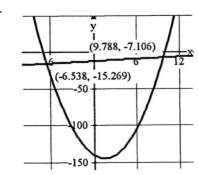

   $x \leq -6.538$ *or* $x \geq 9.788$

   c. solution: all real numbers
   d. no solution
   e. $x < -3.162$ *or* $x > 3.162$
   f. $-3.117 \leq x \leq 2.567$

3. a. (-0.954, -4.908); (3.354, 3.708)
   b. (-1.396, 2.247); (1.396, 2.247);
       (-0.583, -2.581); (0.583, -2.581)
   c. (0.667, 4.667); (3, 3.5)
   d. (2.835, 48.661); (19.138, 16.542)

4. a. $x < 0.667$ *or* $x > 3$

   b. $2.835 \leq x \leq 19.136$

5. a. (M,N); (R,T)
   b. $O < x < P$
   c. $x \leq M$ *or* $x \geq R$

6. a. (S, T); (M, N)
   b. $0 \leq x \leq W$
   c. $x < S$ *or* $x > M$

7. a.

   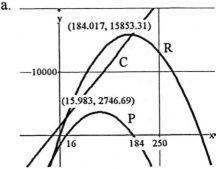

   b. $x = 15.983 \rightarrow y = 2746.69$
       $x = 184.017 \rightarrow y = 15853.31$
   c. $P = -0.51x^2 + 102x - 1500$
   d. see a.
   e. see b.

   f. The x-values for the breakeven pts equal the horizontal intercepts. The profit graph crosses the x-axis directly below the breakeven points.

8. tangent line is $y = 2.4x + 33.8$

   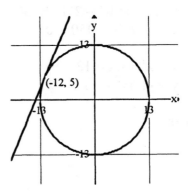

9 – 13. Solutions will vary

Problem Set 7.6

1. a. $(x + 7)(x + 3)$
   b. $x(x - 3)$
   c. $(x - 7)(x + 1)$
   d. $(x + 11)(x - 5)$

2. a. $(x - 4)(x - 1)$
   b. $(x + 9)(x + 4)$
   c. $x(2x + 5)$
   d. $(x - 4)(x - 2)$
   e. $(x - 9)(x+1)$
   f. $(x - 6)(x + 6)$

3. a. $x = 6; x = -5$
   b. $x = -5; x = -7$
   c. $x = 8; x = -2$
   d. $x = 7; x = 3$
   e. $x = 0; x = -15$
   f. $x = 4; x = -4$

4. a. $x = 13; x = -1$
   b. $w = 3; w = 4$
   c. $x = 4; x = -3$
   d. $x = 5; x = 10$
   e. No solution
   f. $m = -6; m = -4$
   g. $x = -\frac{3}{2}; x = \frac{4}{3}$
   h. $x = -16; x = 14$
   i. $x = \frac{5}{3}; x = -\frac{5}{3}$

5. a. $(x - 2)(x +7) = 0$
       $x^2 + 5x - 14 = 0$

   b. Example: $R^2 - 25R + 150 = 0$
       $3R^2 - 75R + 450 = 0$

6. a. $y = x^2 + 4x - 12$
   b. $y = -1.5x^2 - 13.5 - 21$

7. 3 yds.

Problem Set 8.1

1. $\dfrac{7.4}{12.2} = \dfrac{3.1}{TU} = \dfrac{5.2}{UV}$

   $7.4 * TU = 12.2 * 3.1$

   $TU \approx 5.1$ cm

   $7.4 * UV = 5.2 * 12.2$

   $UV \approx 8.6$ cm

2. $DE \approx 2.9$ m

   $DF \approx 5.6$ m

3. $DE \approx 2.9$ cm.

   $AC \approx 4.3$ cm.

4. $\triangle DBC \sim \triangle EBA$

   $\angle D \cong \angle E$ because alternate interior angles of parallel lines are $\cong$.

   $\angle ABE \cong \angle CBD$ because they are vertical angles. $\triangle$ s with two pairs of $\cong$ angles are $\sim$.

5. a. $\triangle ABC \sim \triangle HGI$ since $\angle G = 6°$; the angles are $\cong$

   b. $\dfrac{GI}{5.24} = \dfrac{2.99}{8.05}$; $GI \approx 1.95$ in.

6. $CD \approx 12.9$ ft.

7. $x = 153$ in. $= 12$ ft. 9 in.

8. a. Yes, because $\angle WXV \cong \angle ZXY$ as vertical angles.

   b. $VX \approx 1.29$ in.

   $XZ \approx 0.92$ in.

9. $\triangle ABC \sim \triangle BDC \sim \triangle ADB$

10. a. $\triangle PRT \sim \triangle WYX$

   $m \angle T = 18° = m \angle X$

   $m \angle R = m \angle Y = 90°$

   b. $WX \approx 18.1$ cm

   $NM \approx 13.4$ cm

11. $\dfrac{MK}{3.0} = \dfrac{NK}{10.0} = \dfrac{9.75}{8.5}$

   $MK = \dfrac{3 * 9.75}{8.5} \approx 3.44$ in.

   $NK = \dfrac{10 * 9.75}{8.5} \approx 11.47$ in.

12. $MP \approx 13.9$ m

13. Because of the vertical angles at C, the right triangles are similar.

   $\dfrac{CE}{BC} = \dfrac{DE}{AB}$

14. a. $\triangle ABD \sim \triangle ACE$; $m \angle ABD = m \angle C$; and $m \angle BDA = m \angle E$ because they are corresponding angles of parallel lines cut by a transversal.

   b. $CE \approx 11.8$ m.

15. a. 0.06

   b. ≈ 0.29946

Problem Set 8.2

1. Draw a right triangle with one leg measuring 4.5 cm and the other leg measuring 6.2 cm.
   a. The hypotenuse of the scale drawing measures approximately 7.5 cm → 75 cm.
   b. 36°; 54°

2. 6.5" = $^3/_4$ in.
   x = 9.75 in.

3. P ≈ 98 ft.; A ≈ 570 ft^2

4. $\dfrac{4}{6} \neq \dfrac{8}{11}$;the ratios don't match, you could enlarge to 8 x 12 and then trim a half inch off of each side of the length.

5. 3.81 in. by 6.35 in.
   20.7 in. by 3.45 in.

6. a. AB ≈ 11.6 cm. BC ≈ 6.7
      $m\angle$B = 55°
   b. $m\angle$M =17°,
      NP ≈ 3.0 cm, MP ≈ 10.5 cm.
   c. $m\angle$Q = 41°; SR ≈ $2\dfrac{13}{16}$ in;
      QR ≈ $3\dfrac{3}{16}$ in.
   d. $m\angle$Z = 30°; $m\angle$W = 60°;
      ZY ≈ 4.3 in.
   e. $m\angle$H =138°; IH ≈ 5.1cm.;
      HG ≈ 7.0cm.
   f. $m\angle$D ≈ 33°; $m\angle$E ≈ 77°;
      DE ≈ $1\dfrac{7}{8}$ in.

7. a. Let 55 ft. = 5.5 cm
      KL ≈ 22 ft; JL ≈ 59ft;
      $m\angle$L = 68°

   b. Let 205 yd. = 20.5 cm.
      $m\angle$X = 80°; ZX ≈ 177 yds;
      XY ≈ 139 yds.

8. The units are different.

9. CP ≅ 73 ft.

10. River is approx. 10 m. wide.

11. Approximately 75 yds.

12. Approximately 743 ft.

13. Approx 27990m = 28 km

14. Convert 1 mile to 5280 ft. The slope angle is approx. 4°.

15. slope angle ≈ 2°; horizontal distance ≈ 624 ft.

16. CB is approx. 12.8 m; length of pipe ≈ 44.1 m.

17. 226.2 acres

18. 50 yds; S 36° E

19. total dist. ≈ 8.9 nautical miles
    radio signal distance ≈ 4.75 nautical miles

Problem Set 8.3

1. a. $\sin(x) = \dfrac{9}{15}$ b. $\cos(x) = \dfrac{12}{15}$

   c. $\tan(x) = \dfrac{9}{12}$ d. $\sin(y) = \dfrac{12}{15}$

2. a. – c. Answers will vary depending
       on triangles drawn.

   d. Results will be slightly different
       because it is difficult to measure
       accurately.

3. Solve for hypotenuse.
   $AB = \sqrt{5^2 + 12^2} = 13$

   a. $\sin(A) = \dfrac{5}{13}$ b. $\cos(A) = \dfrac{12}{13}$

   c. $\tan(A) = \dfrac{5}{12}$ d. $\cos(B) = \dfrac{5}{13}$

4. a. $x \approx 6.7$ cm
   b. $x \approx 10.5$ cm

5. a. $QR \approx 3\dfrac{3}{16}$

       $SR \approx 2\dfrac{13}{16}$ in.

   b. $TY \approx 4.3$ in.
       $TW = 5$ in.

6. $\sin 36^\circ = \dfrac{43}{CP}$

      $CP \approx 73.2 \approx 73$ ft.

7. ≈ 10 m

8.   river ≈ 780 ft.

9.

   $5280^2 - 375^2 = x^2$
   $x \approx 5266.7$ ft.
   % slope $= \dfrac{375}{x} \approx 0.071 \approx 7.1\%$

10. ≈ 27990 m ≈ 27.99 km
11. $\approx 66.4^\circ$
12. ≈ 74.6 yds.

13.

   $\tan 31.5^\circ = \dfrac{x}{15}$
   $x \approx 9.19$ ft.
   tree $\approx 9'\,2'' + 5'\,8'' \approx 14'\,10''$

14.a. ≈ 46m; ≈ 31m.
    b. ≈ 1110 m^2

15.

   $\tan(22.5^\circ) = \dfrac{rise}{16} \rightarrow$ rise ≈ 6.6ft.
   $\cos(22.5^\circ) = \dfrac{16}{rafter} \rightarrow$ rafter ≈ 17.3ft.

16. ≈ 44.1m

17.a. Area ≈ 134.3 cm^2
    b. Perimeter ≈ 57.9 cm

18.a. Area ≈ 20.76 in^2
    b. $V \approx 2.89$ ft$^3 \approx 21.6$ gal.

19. ≈ 0.44 mi.

20.a. ≈ 2.165 in.
  b. ≈ 2.289 in.

21.a.

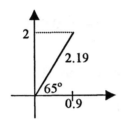

  b. $\sin(65°) \approx 0.91$
       $\cos(65°) \approx 0.42$
       $\tan(65°) \approx 2.14$

  b.   slope $= \dfrac{2}{0.9} \approx 2.22$

  c.   The tangent corresponds to the slope.
       $\dfrac{\text{opposite}}{\text{adjacent}} = \dfrac{\text{rise}}{\text{run}}$ for an angle drawn at the origin.

22.a. $\sin(30°) = 0.5000$
       $\cos(30°) \approx 0.8660$
       $\tan(30°) \approx 0.5774$

  b.   0.7071, 0.7071, 1.0000
  c.   0.9659, 0.2588, 3.7321

  d. $\sin(x)$ is increasing; $\cos(x)$ decreasing, $\tan(x)$ increasing (more rapidly than $\sin(x)$)

  e.   No, in the ratio $\dfrac{\text{opposite}}{\text{hypotenuse}}$, the hypotenuse is always larger, therefore the ratio < 1.

  f.   60°, 45°

Problem Set 8.4

1. a. $\sin(T) = \dfrac{6.2}{9.5}$

   $T = \sin^{-1}\left(\dfrac{6.2}{9.5}\right) \approx 40.7°$

   b. $\cos^{-1}\left(\dfrac{7}{8}\right) = T \rightarrow T \approx 29.0°$

   c. $\tan^{-1}\left(\dfrac{15.7}{8.4}\right) = T \approx 61.9°$

   d. $\tan^{-1}\left(\dfrac{14.5}{8.5}\right) = T \approx 59.6°$

2. a. $H \approx 42.4°$

   b. $W \approx 5.1$ in.

   c. $x \approx 58$ ft.

   d. $G \approx 26.9°$

3. $A \approx 9.5°$
      $x \approx 84.4$ft.
      note: could also use Pyth. Thm.

4.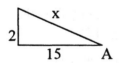

   1 chain = 66 ft.

   $x = \sqrt{15^2 + 2^2} = 15.13$ chains

   ≈ 998.76 ft. ≈ 999 ft.

   $\tan(A) = \dfrac{2}{15}; \; A \approx 7.6°$

5. a. $4.76°$
   b. $\approx 4.76°; \; \approx 85.24°$

6. ≈ 12.5 ft

7. $\sin(A) = \dfrac{375}{5280} \rightarrow \; A \approx 4.07°$

8.   horizontal distance ≈ 6.2 ft.
      diff. in elevation ≈ 516.96 ft.

9. $\tan(2°) \approx 0.0349 \approx 3.49\%$

10. $R \approx 3.09$ in
      1 stroke $\approx 9.71"$

11. $\angle B \approx 30.3°$

12. a. $124°32'$
    b. $x \approx 170.1$ ft.

13. $A \approx 18.4°$

14. a. $\sin(20°) \approx 0.3420,$
      $\cos(28°) \approx 0.8829,$
      $\tan(36°) \approx 0.7265$

    b. $38°, 10°, 22°$
    c. $25°$

15. a. $11.537°, 56.633°, 45.00°$
    b. $\approx 0.391, \approx 0.174, \approx 0.259$

Problem Set 9.1

1. a. f(2) = 6, f(4) = -1, f(-1) = 3
   b. g(2) = -3, g(4) = 6, g(-1) = 3

2. a. $f(3) = 3^2 - 5 = 4$

   b. g(-5) = -3

   c. $h(15) = \sqrt{2*15-5} = 5$

   d. $F(3) = \dfrac{3}{4}$

   e. $G(-2) = 3(-2)^3 - 7(-2)^2 + 5 = -47$

   f. H(7) = 6

3. a. $S(3.2) \approx 184.98 \text{ ft}^2$
   b. $S(5.1) \approx 355.69 \text{ ft}^2$

4. a. P(10) = 17.6
      In 1990 the population of Little Rock, AK was 176,000.
   b. P(17) = population in 1997.

5. a. Ian wears a size $10\dfrac{1}{2}$ shoe.

   b. The shoe size Penny wears.

6. a. The cost of driving 1,000 miles is $650.
   b. The cost of driving 450 miles.

7. a. $F(2) = \dfrac{4}{3}$ b. $F(4) = \dfrac{16}{5}$

   c. F(0) = 0   d. $F(t) = \dfrac{t^2}{t+1}$

   e. $F(m + 1) = \dfrac{(m+1)^2}{m+2}$

   f. F(-1) = undefined

8. a. 12 = 2x + 4.5
      7.5 = 2x → x = 3.75

   b. x ≈ 1.3699; x ≈ -1.7033

c. $x = \dfrac{13}{5}$

d. x = 3

e. x = 1; x = -11

f. x = 1.5 *or* x = -6.5

9. a. f(3) = 5; f(-5) = 0
      f(0) = 4; f(-6) = 2

   b. g(-3) = 2; g(4) = 4
      g(0) = -3; g(-5) = -4
      g(8) = 4

10. domain: x is all real numbers
    range: $y \geq 0$

    domain for g(x): $x \geq -5$
    range for g(x): $-4 \leq y \leq 4$

11. a. $h(-4) \cong -2.5$; h(-1) = 3; h(2) = 3
    b. k(2) = -4; k(4) = 0; k(-1) = -1

12. domain of h(x): all reals
    range: $y \leq 3$

    domain of k(x): all reals
    range: $y \geq -4$

13. a. domain: all reals
       range: all reals

    b. domain of g(x) : all reals
       range: $g \geq -135$

    c. domain h(t): all reals
       range: $h \geq 0$

    d. domain: all reals
       range: $k \leq -10$

    e. domain F(t): $t \geq 5$
       range: $F \geq 0$

    f. domain G(x): all reals
       range: $G \geq -81$
       (hint: find vertex)

14. a. domain: all reals
       range: $F(x) \geq 0$

b.   domain: $x \leq 0$
range: $K(x) \geq 0$

c.   domain: all reals except 2
range: all reals except 0

d.   domain: $t \geq 0$
range: $H(t) > 0$

e.   domain: all reals
range: $g(t) \geq 0$

15.a.   Yes, each person has only one birthday as an output.

b.   Yes, see a. Note: if the input was the date and the output was the name of the person it would not be a function.

16.a.   Not a function. People could have more than one phone number listed.
b.   Maybe. Yes, if each person has only one shoe size.
c.   No, for February it depends on leap year.

17.   Volume $= (3 - 2x)(2 - 2x) \, x$
$0 < x < 1$ in. Otherwise the width would be 0 or negative.

Problem Set 9.2

1.a. Yes, assuming rate of royalties does not change with the number sold. Input of inverse = amount earned in royalties, $R \geq 0$; output of inverse = number of CD's valid, $n \geq 0$.

b. No, more that one player can have the same percentage of hits.

c. No, relationship depends on changes in speed and time; therefore, you could have more than one method of burning off a certain number of calories.

2. a. No; eventually you would reach a cruising speed.

b. Yes; input = area; output = width of square

c. Yes; if cost is charged in some uniform way. input = cost, output = Kilowatt hrs. used
or
No; if there are variable rates, like evening hours are cheaper than day hours, you could get the same bill with different usage.

d. No; many students share the same shoe size.

e. No; Rick may change weight during the year; Rick may weigh the same for several years making the inverse not a function.

f. Yes; input = number of bacteria, output = time in min.

3. a. g(x) does have an inverse that is a function
h(x) does not; k(x) does

b.

x	$g^{-1}(x)$
-2.5	-5
0	0
2.5	5
5	10
7.5	15

x	$k^{-1}(x)$
-1	$^1/_3$
0	1
1	3
2	9
3	27

c. $g(5) = 2.5$; $g^{-1}(5) = 10$
$h(-1) = 6$
$k(3) = 1$; $k^{-1}(3) = 27$

4. a. Population in 1980.
b. Population in 2000.
c. The year when the population equals 250,000.

5. a. Cost of a house with 800 sq. feet.
b. Square footage of a house that costs $100, 000.
c. Square footage of a house that costs $200,000.

6. a. Cost of a 50 mile trip.
b. Number of miles driven if the cost is $100.
c. Cost of a 100 mile trip.

7. a. Number of centenarians living in Florida in 2050.
b. The year that 2100 centenarians are living in Florida
c. The year that 4000 centenarians are living in Florida.

8. a. The velocity of the rock after 10 seconds.
b. The time when the velocity will equal 21 f/s
c. The velocity of the rock after 60 seconds.

9. a. $x = y + 2 \;\rightarrow\; y = x - 2$

   b. $x = 3y - 4 \;\rightarrow\; y = \dfrac{x + 4}{3}$

   c. $x = \dfrac{y}{2} + 5 \;\rightarrow\; y = 2x - 10$

   d. $x = y + \dfrac{5}{2} \;\rightarrow\; y = x - 2.5$

10. a. $y = x^3$ b. $y = \sqrt[3]{x + 4}$

    c. $y = \dfrac{x}{3} - 6$ d. $y = x^5 - 1$

11.   domain of f^{-1} = all positive real numbers
      range of f^{-1} = all real numbers
      Since the values of the function reverse, the domain and range also interchange.

12. a.

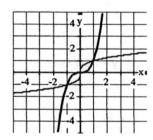

The parabola does not pass the horizontal line test.

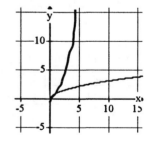

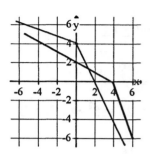

   b. $f(-1) \approx -1;\; f(4) \approx 2;\; g(2) \approx 1;$
        $h(4) \approx 3;\; h(9) \approx 4;\; k(4) \approx -4$

   c. $f^{-1}(-1) \approx -1;\; f^{-1}(1.5) \approx 1.5;$
        $h^{-1}(2) \approx 3;\; h^{-1}(3) \approx 3.5;\; k^{-1}(-4) \approx 4$

13. a.   No. Reverse x and y: $x = y^2$
         solve: $y = \pm\sqrt{x}$;
         You get two outputs.

   b.   Yes $x = \sqrt{y} \;\rightarrow\; y = x^2$
        For each input there is only one output.

   c.   domain of x^2: all reals
        range of x^2: $y \ge 0$, all positive reals

   d.   domain of $\sqrt{x}$: $x \ge 0$ or all positive numbers
        range of $\sqrt{x}$: $y \ge 0$ all positive numbers

   e.

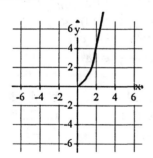

   f.   The graph equals the positive side of $y = x^2$.

<u>Problem Set 10.1</u>

1. a. $6^x = 36$
         $6^x = 6^2 \to \quad x = 2$

   b. $5^x = 25 = 5^2 \to \quad x = 2$

   c. $4^x = \dfrac{1}{16}$
         $4^x = 4^{-2} \to \quad x = -2$

   d. $(\frac{1}{2})^x = 8$
         $(\frac{1}{2})^x = (\frac{1}{2})^{-3} \to \quad x = -3$

   e. $5^x = 1 \to \quad x = 0$

   f. $10^x = 1000 = 10^3 \to x = 3$

2. a. $x = 3$ b. $x = 4$
   c. $x = 0$ d. $x = -2x + 3$
   e. $x = -3$ f. $x = 4$
   g. $x = 2$

3. a. $\log_3(100) \to 3^x = 100$
         $3^4 = 81$ *and* $3^5 = 243$
         $3^4 < 3^x < 3^5$
         $4 < x < 5$

   b. $\log_3(0.01) \to 3^x = 0.01$
         $3^{-5} = 0.004$ *and* $3^{-4} = 0.012$
         $0.004 < 0.01 < 0.012$
         $3^{-5} < 3^x < 3^{-4}$
         $-5 < x < -4$

4. a.   1.24   b.   4.12
   c.   0.602   d.   2.30
   e.   2.41   f.   –2

5. a. $2^5 = x$ b. $\log_4(16) = m$

   c. $x^{\frac{1}{2}} = 5$ d. $4^{-\frac{1}{2}} = x^2$

   e. $e^4 = -x$

   f. $\log(12000) = -m$

6. a. $3^4 = x \to \quad x = 81$

   b. $2^t = 2^4 \to \quad t = 4$

c. $b^{\frac{1}{2}} = 3 \to \sqrt{b} = 3 \to \quad b = 9$

d. $x \approx -1.754$

e. $x \approx -548.3$

f. $k = 4$

g. $k \approx 2.845$

7. a. $x = 49$ b. $t = 2.868$
   c. $n = -2$ d. $x \approx 4.055$
   e. $x \approx 1.447$ f. $x \approx -0.9960$

8.

x	$y = \log_5(x)$
$\frac{1}{5}$	-1
1	0
5	1
25	2
10	1.431
60	2.544
200	3.292

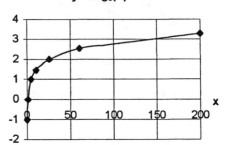

$y = \log_5(x)$

9.

	(a) equivalent	(b) inverse
$y = \log_2(x)$	$x = 2^y$	$y = 2^x$
$y = \log(x)$	$x = 10^y$	$y = 10^x$
$y = \ln(x)$	$x = e^y$	$y = e^x$

10. a.   False; $2^{\frac{1}{3}} \neq \dfrac{1}{8}$

   b.   True

   c.   False; $2^{-3} = \dfrac{1}{8} \neq -8$

   d.   False; $5^{25} \neq 2$

11.a. $y = x$

b. $y = x$

c. $y = x$; domain is all real numbers when $x < 0$, b^x will still be positive and $\log_b$ (positive) is defined.

d. $y = x$ where $x > 0$

e. $y = x$ where $x > 0$

f. $y = x$ where $x > 0$
   x must be greater than 0 because $\log_b(x)$ is not defined for negative values of x.

12.a. $0 < m < 1000$

b. $0 < P \le 0.0183$

13.a. $\ln(4) = 1.386$
   $\log(4) = 0.602$
   $\ln(4) > \log(4)$
   since $e < 10$ need a greater exponent for value to equal 4.

b. $\ln(2) > \log(2)$
   for $e^x = 10^y = 2$; $x > y$

c. $\ln(1) = \log(1)$
   since $e^0 = 10^0 = 1$

d. $e^x = 10^y = \frac{1}{2} \rightarrow x < y$
   $\therefore \ln(0.5) < \log(0.5)$

e. $\ln(0.1) < \log(0.1)$

f. $\ln(x) < \log(x)$ when $0 < x < 1$

14.a. $2^x = 5^y = 10 \rightarrow x > y$
   $\therefore \log_2(10) > \log_5(10)$

b. $2^x = 5^y = 6 \rightarrow x > y$
   $\therefore \log_2(6) > \log_5(6)$

c. $\log_2(1) = \log_5(1)$
   since $2^0 = 5^0 = 1$

d. $2^x = 5^y = \frac{1}{2} \rightarrow x > y$
   $\therefore \log_2(\frac{1}{2}) < \log_5(\frac{1}{2})$

e. $\log_2(^1/_5) < \log_5(^1/_5)$

f. $\log_2(x) < \log_5(x)$ when $0 < x < 1$

15.a. bicarbonate = 26 mEq/l
   pH = 7.4

b. Yes, carbonic acid is greater than 2.

Problem Set 10.2

1. a. Exponential   b. Quadratic
   c. Linear   d. Other
   e. Logarithmic   f. Power

2. a. ≈ 3.32 b. ≈ 0.75
   c. ≈ 158.61 d. ≈ 0.74

3. a. ≈ 3.49

   b. $5^{-1} = 2x - 1$

   $x = \dfrac{0.2 + 1}{2} \approx 0.6$

   c. $x \approx 21.54$

   d. $3^3 = 2m - 7$
   $2m = 27 + 7 \rightarrow m = 17$

   e. No solution

   f. $\log_2\left(\dfrac{x}{x-1}\right) = -2$

   $2^{-2} = \dfrac{x}{x-1} \rightarrow \dfrac{1}{4} = \dfrac{x}{x-1}$

   $x - 1 = 4x \rightarrow -1 = 3x$

   $x = -\dfrac{1}{3} \approx -0.33$

   but $x > 0$, therefore no solution

   g. $n \approx 2.67$

   h. $\log(3x^2 + 15x) = 1$
   $10' = 3x^2 + 15x$
   $0 = 3x^2 + 15x - 10$

   $x = \dfrac{-15 \pm \sqrt{15^2 - 4(3)(-10)}}{6}$

   $x \approx 0.60$ (x must be positive)

4. a. ≈ 3.71 b. ≈ 0.30
   c. ≈ 1.84 d. ≈ -0.60
   e. ≈ 14.21 f. ≈ 6.99

5. a. ≈ 3.25 b. ≈ 3.18
   c. ≈ -3.49 d. ≈ 8.45
   e. ≈ 14.21 f. ≈ 6.99

6. a. $R = \log(I) \rightarrow I = 10^R$

   b. $P = P_0 e^{rt} \rightarrow \dfrac{P}{P_0} = e^{rt}$

   $\ln\left(\dfrac{P}{P_0}\right) = rt$

   $t = \dfrac{1}{r}\ln\left(\dfrac{P}{P_0}\right)$

   c. $M = \ln\left(\dfrac{x}{x-1}\right)$

   $e^M = \dfrac{x}{x-1}$

   $e^M * x - e^M = x$

   $e^M * x - x = e^M$

   $x(e^M - 1) = e^M$

   $x = \dfrac{e^M}{e^M - 1}$

   d. $\dfrac{L}{10} = \log\left(\dfrac{I}{I_o}\right)$

   $10^{\frac{L}{10}} = \dfrac{I}{I_0} \rightarrow I = I_0 * 10^{\frac{L}{10}}$

   e. $\left(\dfrac{A}{A_0}\right) = e^{-kt}$

   $\ln\left(\dfrac{A}{A_0}\right) = -kt$

   $kt = -\ln\left(\dfrac{A}{A_0}\right) \rightarrow k = -\dfrac{1}{t}\ln\left(\dfrac{A}{A_0}\right)$

   f. $2 = \left(1 + \dfrac{r}{n}\right)^{nt}$

   $\log(2) = nt * \log\left(1 + \dfrac{r}{n}\right)$

   $t = \dfrac{\log(2)}{n * \log\left(1 + \dfrac{r}{n}\right)}$

g. $P_0 * 2^{1/4} = P_0 e^{kt}$

$2^{\left(\frac{t}{4}\right)} = e^{kt}$

$\left(\frac{t}{4}\right)\ln(2) = kt$

$k = \left(\frac{t}{4}\right) * \frac{\ln(2)}{t}$

$k = \frac{\ln(2)}{4} \approx 0.1733;\ t \neq 0$

7. $\log_b(x) = u$ and $\log_b(y) = w$ then
     $b^u = x$ and $b^w = y$

$\frac{x}{y} = \frac{b^u}{b^w} = b^{u-w}$

$\log_b\left(\frac{x}{y}\right) = u - w$

$\log_b\left(\frac{x}{y}\right) = \log_b(x) - \log_b(y)$

8.   carbonic acid $= \dfrac{\text{bicarbonate}}{10^{(pH-6.1)}}$

9. ≈ 11 years
10. ≈ 10.7 yrs.

11.a.

x	$^1/_5$	1	5	25	125
y	-1	0	1	2	3

$y = \log_5(x)$

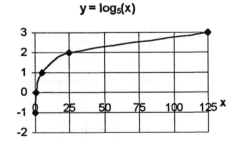

b. ≈ 2.3 c. $5^y = 50$

d. ≈ 2.43 e. $y = \dfrac{\log(x)}{\log(b)}$

12.a. $B_1 = \$1233.55$
       (Banks round down)
       $B_5 = \$6977.00$
   b. ≈ 158 months ≈ 13 yrs 2 months

13. $t \approx 50.23$ hours

14.a. $I = 3, 162, 277, 660$
   b. Based on OSHA reports, if you
       will be listening to the noise:
          for 1 hour →105 dB;
          for 15 min. →115 dB;
          for 8 hrs. → 90 dB.
   c. No

15.a. $x = -0.6$ b. $x \approx 2.773$

   c. $x = \dfrac{1}{15}$ d. $x \approx 7.397$

   e. $x \approx -5.864$ f. $x = 200$

   g. Not defined   h. $x \approx \pm 1.015$

16.a. $A = \$2.30;$ c.
       $A = \$2.67$
   b. $A = \$1.81$

t	A
-5	$1.64
-4	$1.72
-3	$1.81
-2	$1.90
-1	$1.99
0	$2.09
1	$2.19
2	$2.30
3	$2.42
4	$2.54
5	$2.67

d.

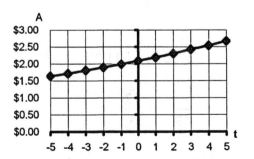

17. $r \approx 7.2\%$

Chapter Review   7 – 10

1. – 5. Answers will vary.

6.   Let $Y_1 = -3x^2 - 5x + 10$;
         $Y_2 = 0.5x + 4$ *and*
         $Y_3 = Y_1 - Y_2$
         Set table to Auto, ΔTbl = 1
         Look in the Y_3 column for values
         closest to 0.
         Reset ΔTbl. = 0.1
         with Tbl Start = values of x
         that gave Y_3 closest to 0.

7.   Let $Y_1 = -3x^2 - 5x + 10$
         $Y_2 = 0.5x + 4$
         Set window to standard screen.
         Use intersection program to find
         both points of intersection.
         Solution equals the x-values.

8.   Since the equation is quadratic,
         rearrange terms so that equation
         is in standard form $3x^2 + 5.5x - 6 = 0$. Use the quadratic formula
         to solve for x.

9.

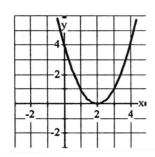

10. $y = -3x + 39$
11. $y = 8.125x + 131.25$

12. $y = \dfrac{^-5}{3}x + \dfrac{17}{3}$

13.a.   Line   b.   Line
    c.   Parabola   d.   V
    e.   Other   f.   Line
    g.   Parabola

14. a. $y = x$

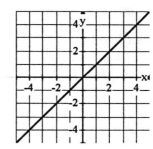

b. $y = -5x + 100$

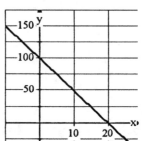

c. $x = -4$

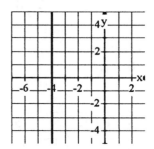

d. $y = 5.4$

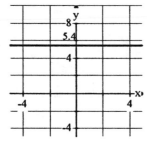

e. $4.6x + 0.25y = 24$

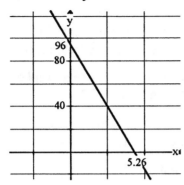

15. $x \approx -0.2636353$

16.a. Line   b. Circle
  c. Parabola   d. Exponential
  e. Other   f. Parabola
  g. Exponential   h. Other
  i. Parabola

17.a. Parabola opening up with
    vertical-intercept (0,15)
    window: [-10, 10]; [-5, 50]

  b. Wide parabola opening down
    with vertex at (0,0)
    window: [-10, 10] [-5, 50]

  c. Wide parabola opening down,
    vertical-intercept at (0, 20)
    window: [-15, 15] [-5, 25]

  d. Parabola opening up through (0,0)
    (15, 0).
    window: [-10, 20] [-80, 20]

  e. Parabola facing down through
    (0,0) and approximately (≈ 5, 0)
    window: [-5, 10] [-10, 80]

  f. Parabola facing down with
    vertical-intercept at (0, -2300)
    window: [-20, 100] [-3000, 750]

18. $(19\frac{1}{3}, -5\frac{1}{3})$

19.a. $x \approx -91.0$

b. $r \approx 0.0371$
c. $x \approx \pm 1.45$
d. $x = -65$

20. Answers will vary:
    $y = -7x;\ y = 2 - 7x;\ y = -1 - 7x$

21. $y = -0.25x + 4.5$

22. Answers will vary.
    $y = 1.5x;\ y = 1.5x + 2;\ y = 1.5x - 1$

23. $(\approx 1.82,\ \approx 11.35);$
    $(\approx -17.82,\ \approx 50.65)$

24.a. (-2, 6); (4, -6)

  b. (-18.362, 116.808);
    (1.362, 18.192)

  c. (-3.161, -0.081);
    (1.961, 2.481)

  d. (0.368, 6.386); (16.299, -20.164)

25.a. $x < -16.55\ or\ x > 3.22$
  b. no solution

26.a. $k = 0;\ \dfrac{1}{3}$

  b. $t = \dfrac{5}{2} = 2.5;\ t = \dfrac{1}{2} = 0.5$

  c. $T = \dfrac{^-6}{2M - 7}\ or\ \dfrac{6}{7 - 2M}$

  d. $k = \dfrac{1 \pm \sqrt{1 - 12m}}{6}$

27.a. $x > -2$
  b. $x \leq 1$
  c. $x \leq -5\ or\ x \geq 10$
  d. $-2 < x < 3$

28.a. 9   b. 9   c. -8
  d. $\dfrac{1}{125}$ e. 9

29. 188.04

30. ≈ -95.358

31. $5\sqrt{w}$

32. $50x^{-1/6}$

33. No

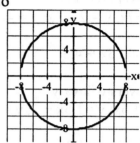

34. For these problems assume $\angle A$ is across from the shortest side and is the smallest angle.

 a. $c \approx 90.9$ cm
 $\angle A \approx 27.7°$
 $\angle B \approx 62.3°$

 b. $\angle B = 66°$
 $b \approx 319$ mm
 $c \approx 349$ mm

 c. $\angle A \approx 46.0°$
 $\angle B \approx 44.0°$
 $b \approx 2.19$ m

 d. $\angle A = 25°$
 $a \approx 5$ in. (5.1 in)
 $b \approx 11$ in.

35. ≈ 45 ft.
36. $h = 6m$

37. $h \approx 43.06$ cm
 $x \approx 27.33$ cm
 $y \approx 13.59$ cm
 $A \approx 2826.5$ cm^2

38. $x \approx 4.887$ (4.9 mi)
 S 18.3°E

39. Perimeter ≈ 40 cm
 Area ≈ 110 cm^2

40. $r \approx 0.2478$ ft ≈ 2.97 in.

41.a. Domain: x = all real numbers
 Range: f(x) = all real numbers

 b. Domain: x = all real numbers
 Range: $g(x) \geq 12$

 c. Domain: t is any real number
 Range: $h(t) \geq 0$

 d. $t \neq 3$; $k(t) \neq 0$

 e. $x \geq -1$; $F(x) \geq 0$

42.a. $x = 1.3$
 b. $x = -2.5$; $x = 1$
 c. $x = -1$

43.a. $x = -5$; $x = 1$
 b. 45

44.a. $f^{-1}(x) = \dfrac{x-9}{2}$

 b. $f^{-1}(x) = \sqrt[3]{x-6}$
 c. $f^{-1}(x) = 2x + 26$

45.a. $B^{-1}(200)$= number of kilowatt hours used when the bill is $200.

 b. B(200) = amount of monthly bill when 200 kilowatt hours of electricity are used during the month.

 c. $B^{-1}(165)$ = number of kilowatt hours used when the bill is $165

46.a. T(50) = time (in minutes) to travel 50 miles.

 b. $T^{-1}(100)$ = distance in miles traveled in 100 minutes.

c. $T^{-1}(117) =$ distance in miles traveled in 117 minutes.

47.a. $y(32) =$ temperature in Fahrenheit at 32°C

  b. $y^{-1}(32) =$ temperature in Celsius at 32°F

  c. $y(-40) =$ temperature in Fahrenheit at –40°C

48. $F(-1) = 2;\ F^{-1}(4) = -5;\ G(-2) = 5;$ $G^{-1}(8) = 6;\ G^{-1}(-2) = 0$

49. Domain: $x \geq -5$; Range: $F(x) \leq 4$

50.a. Exponential   b. Linear
  c. Neither   d. Exponential

51.a. $C = 1.25t + 5.75$

  b. $B = 7 * 1.2^n$

  c. $P = 1458 * 3^{-m}$ or $1458 * \left(\dfrac{1}{3}\right)^m$

  d. $B = 3n - 86$

52.a. $y = 5^x$; Increasing exponential curve through (1,5).

  b. Decreasing exponential curve through (-1,4) and (1, 0.25)

  c. Decreasing exponential curve through (-1, 12) and (1, 3)

  d. Decreasing exponential curve through (-1, 3) and (1, $^1/_3$)

  e. Increasing exponential curve through (0, 3.7) and (1, 29.6)

  f. Increasing exponential curve through (1, e)

53. $N \approx 42$ days

54.a. $N = 5 * 2^t$
  b. $D = 1.25P - 0.75$

55. $\log_2(3x) = \log_2(x) + \log_2(3)$
$\log_2(8x) = \log_2(x) + 3$
$\log_5\left(\dfrac{1}{5}\right) = \log_5(1) - \log_5(5)$
$\log\left(2^{x/5}\right) = \dfrac{x\log(2)}{5}$

56.a. $2^5 = x$
  b. $x^2 = 64$
  c. $10^2 = 2x + 50$
  d. $(x^2)^3 = 64$

57.a. 3   b. 2   c. –2
  d. 5   e. 0   f. 4
  g. 4

58.a. $t \approx 33.12$
  b. $x = {}^{17}/_5 = 3.4$
  c. $x = 16$
  d. $m = 732.5$
  e. $k \approx 49.74$
  f. $x = 5$

59.a. $x = e^{1.2} \approx 3.32$
  b. $x \approx 0.75$
  c. $x \approx 158.61$
  d. $x \approx 0.74$

60.a. $x \approx 1.636$
  b. $x \approx -0.696; x \approx 1.196$
  c. $x \approx 6.604$
  d. $x \approx -2.191; x \approx 2.191$
  e. $x \approx 20.781$
  f. $x \approx 1.405$

61.a. ≈ 55 decibels
  b. $P = 10^{-6}$
  c. $P = 0.01$ watts/cm^2
    10,000 times higher

62.a. pH = 8
  b. pH = 4.4 is acidic

c. $H^+ \approx 2.51 * 10^{-6}$
approximately 16 times higher

d. ≈ 0.005
approximately 126 times higher

63.a. Increasing exponential curve through (0,2) and (1, 10.2) with horizontal asymptote at $y = 0$.

b. Increasing logarithmic curve with vertical asymptote at $x = 0$ through (1,0) and (e, 1)

c. Decreasing exponential curve through (-1, 300) and (1, 75) with horizontal asymptote at $y = 0$.

d. Increasing logarithmic curve with vertical asymptote at $x = 0$ through (1,0) and (10, 1)

e. Line through (0,14) with m = -7.

f. Decreasing exponential curve through (-1, 12) and (0,4).

64.a. $x \approx 3.75$
b. $t \approx 0.30$
c. $x \approx 2.07$
d. $x \approx 2.51$
e. $x \approx 7.00$
f. $t \approx 0.49$